T0140380

Advances in Experimental Medicine and Biology

Clinical and Experimental Biomedicine

Volume 1096

Subseries Editor
Mieczyslaw Pokorski

More information about this series at http://www.springer.com/series/16003

Mieczyslaw Pokorski

Editor

Rehabilitation Science in Context

 Springer

Editor
Mieczyslaw Pokorski
Opole Medical School
Opole, Poland

ISSN 0065-2598   ISSN 2214-8019   (electronic)
Advances in Experimental Medicine and Biology
ISSN 2523-3769   ISSN 2523-3777   (electronic)
Clinical and Experimental Biomedicine
ISBN 978-3-030-07086-1   ISBN 978-3-319-95708-1   (eBook)
https://doi.org/10.1007/978-3-319-95708-1

This Springer imprint is published by the registered company Springer Nature Switzerland AG
The registered company address is: Gewerbestrasse 11, 6330 Cham, Switzerland

Contents

Adv Exp Med Biol - Clinical and Experimental Biomedicine (2018) 1: 1–9
https://doi.org/10.1007/5584_2018_187
© Springer International Publishing AG, part of Springer Nature 2018
Published online: 29 March 2018

Improvement in Gait Pattern After Knee Arthroplasty Followed by Proprioceptive Neuromuscular Facilitation Physiotherapy

Joanna Jaczewska-Bogacka and Artur Stolarczyk

Abstract

The aim of the study was to assess the influence of a physiotherapy program based on proprioceptive neuromuscular facilitation (PNF) on kinematic gait pattern after total knee arthroplasty. This comparative study included two groups of patients qualified for total surgical knee joint replacement due to osteoarthritis: a study group and a control group, either consisting of 28 patients of a matched age range of 55–90 years. Following surgery, 4 days after standard postoperative rehabilitation, the study patients were subjected to a 3-week-long therapist-assisted rehabilitation based on PNF principles (10 sessions of 75 min each), whereas control patients were discharged with instructions on how to exercise in the home setting. The outcome consisted of spatial-temporal gait parameters that were assessed at three time points: a day before surgery and then 1 month and 6 months after. The findings were that PNF caused substantial, sustained improvements in gait kinematics, shortening the stance phases, gait cycle duration, and double support phase and prolonging swing phase velocity, gait velocity, cadence, step length, and gait cycle length. Also, postsurgical pain was evidently less. We conclude that the individually tailored PNF rehabilitation program is superior compared to a standard recommendation of home-based physiotherapy in terms of improving gait kinematic pattern as well as psychological aspects related to pain perception in patients after knee arthroplasty.

Keywords

Gait kinematics · Gait pattern · Joint replacement · Knee arthroplasty · Physiotherapy · Proprioceptive neuromuscular facilitation · Rehabilitation

1 Introduction

It is estimated that there were approximately 1.4 M knee arthroplasty surgeries carried out around the world in 2015 (Szöts et al. 2015). Studies show that about 20% postoperative arthroplasty patients still complain about chronic knee joint pain (Beswick et al. 2012) and 15% of patients suffer from pain, despite the absence of any underlying reasons in radiological joint images (Lee et al. 2015). Pathological gait pattern after surgery also is observed in patients who do

J. Jaczewska-Bogacka (✉)
Lekmed Medical Center, Warsaw Medical University, Warszawa, Poland
e-mail: jaczewska.joanna@op.pl

A. Stolarczyk
Department of Clinical Rehabilitation, Warsaw Medical University, Warsaw, Poland

not experience pain, which may result from the inability to adjust to new conditions (Stan and Orban 2014). It is necessary to improve therapeutic procedures after knee arthroplasty to maximize the effectiveness of surgery and improve the patients' quality of life.

The aim of the present study was to assess the influence of a physiotherapy program based on proprioceptive neuromuscular facilitation, used in total knee arthroplasty, on the patients' gait pattern in 1 month and 6 months after surgery compared with standard instructions on how to exercise at home.

2 Methods

2.1 Study Organization

The study was approved by the Bioethics Committee of Warsaw Medical University in Poland (approval no. KB/123/2012). The study included patients qualified for total surgical knee arthroplasty due to osteoarthritis, treated in the Orthopedics and Rehabilitation Clinic of the Second Medical Department of Warsaw Medical University in Warsaw, Poland, in 2013–2014.

Patients were divided into a control (23 women and 5 men, age 68.1 ± 6.9, range 55–80 years, mean body mass index (BMI) 31.5 ± 3.8 kg/m^2) and study groups (19 women and 9 men, age 68.7 ± 8.8, range 55–90 years, BMI 31.7 ± 5.6 kg/m^2). Patients of either group had a lower limb operated on and had a different rehabilitation program. All the patients had had knee joint pain before surgery for more than 3 years, and they had no exercise program before surgery. Patients of the control group were recruited in 2013. These patients performed two types of recommended post-surgery exercise at home (active knee flexion and extension and isometric quadriceps contraction). During the following year, recruitment for patients of the study group took place. These patients had home-based exercise akin to that performed by

the control patients above outlined plus individual post-surgery exercise program, based on proprioceptive neuromuscular facilitation. All the patients were examined three times: 1 day before surgery, 1 month after surgery, and 6 months after surgery.

Knee arthroplasty consisted of the implantation of a posterior-stabilized Zimmer Nexgen knee endoprosthesis. Surgical procedures were always performed by the same surgeon under the general same conditions.

2.2 Rehabilitation Program

In the first week after surgery, patients from both study and control groups underwent an early postoperative rehabilitation. Physiotherapy started 24 h after surgery and included four 40-min sessions in the patient's room. The sessions consisted of breathing exercises, exercises improving blood circulation, isometric exercises, knee joint mobilization, exercises in the standing posture, proper walking on crutches, and cooling the joint with cold compresses. After 4 days of postoperative rehabilitation, the patient was discharged and recommended to continue the same kind of exercise with increasing intensity and number of repetitions.

After further 4 days of exercises in the home setting, the study patients participated in 3-week-long proprioceptive neuromuscular facilitation physiotherapy. The program included ten meetings with a physiotherapist (three times a week) of 75-min duration each. Figure 1 illustrates some of the interventions used. A physiotherapy session started with reducing pain and swelling, relaxing tense muscles, and increasing mobility (techniques, hold-relax, contract-relax, and rhythmic initiation). Then, the core muscles were activated to improve trunk and lower limb stability (techniques, stabilizing reversals and rhythmic stabilization). Also, techniques increasing muscle strength and endurance and improving coordination   (combination   of   isotonic

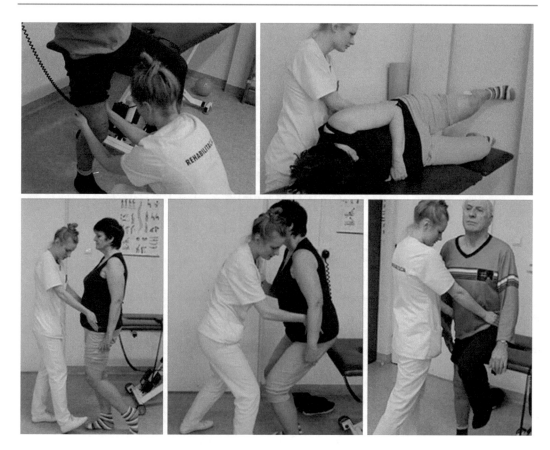

Fig. 1 Examples of proprioceptive neuromuscular facilitation (PNF) treatment

contractions and dynamic reversals) and proprioception and balance training were used in the session. The ultimate goal was to restore the proper gait pattern, so that at the end of a session, different gait phases were directly facilitated. Additional interventions included patella mobilization, scar mobilization, and hands-off balance exercises.

2.3   Rehabilitation Outcomes

BTS Smart system (BTS Bioengineering Corp., Brooklyn, NY), a high-precision optoelectronic system, equipped with infrared illuminators, for the biomechanical motion analysis, was used to assess the patients' gait pattern as they walked a distance of 10 m. The system enables the objective, quantitative analysis of temporospatial parameters. Three sequential trials were performed to calculate the mean results. The trial results were compared against average results provided by the BTS system database (Table 1).

2.4   Statistical Analysis

Differences between the study and control patients were statistically assessed with Student's t-test and the Mann-Whitney U test. Results within a group were compared using the one-way ANOVA with repeated measures or its nonparametric alternative, the Friedman test. Additionally, lower and upper limits of 95% confidence intervals for variances were calculated.

Table 1 Gait pattern changes after knee arthroplasty

Gait pattern	Control patients[a]			Study patients[b]			Healthy adults 65 + years old[c]
	Before	Surgery After 1 month	After 6 months	Before	Surgery After 1 month	After 6 months	
Stance phase – operated limb (s)	1.08 ± 0.36	1.08 ± 0.62	1.03 ± 0.45	1.03 ± 0.34	0.80 ± 0.14 †	0.72 ± 0.11 †*^	0.63 ± 0.02 right
Stance phase – non-operated limb (s)	1.11 ± 0.38	1.10 ± 0.77	1.03 ± 0.50	1.06 ± 0.35	0.82 ± 0.14 †*	0.72 ± 0.09 †*^	0.63 ± 0.04 left
Stance phase (%)	67.80 ± 6.60	67.90 ± 5.30	67.90 ± 5.30	67.70 ± 7.10	62.70 ± 3.80 †*	61.90 ± 3.80 †*	59.60 ± 1.20
Swing phase – operated limb (s)	0.49 ± 0.07	0.47 ± 0.09	0.46 ± 0.09	0.46 ± 0.07	0.47 ± 0.06	0.44 ± 0.05	0.43 ± 0.02 right
Swing phase non-operated limb (s)	0.45 ± 0.07	0.46 ± 0.07	0.47 ± 0.07	0.43 ± 0.06	0.47 ± 0.05	0.45 ± 0.05	0.43 ± 0.02 left
Gait cycle duration (s)	1.56 ± 0.38	1.56 ± 0.81	1.50 ± 0.53	1.49 ± 0.34	1.27 ± 0.17 †*	1.16 ± 0.14 †*^	1.06 ± 0.03
Double-limb support (s)	19.10 ± 6.90	17.50 ± 5.00	18.20 ± 6.30	19.90 ± 9.30	13.30 ± 3.40 †*	11.40 ± 3.90 †*^	–
Swing phase velocity (m/s)	1.38 ± 0.41	1.59 ± 0.45	1.66 ± 0.42 *	1.44 ± 0.41	1.76 ± 0.37 *	2.04 ± 0.51 †*^	3.30 ± 0.14
Gait velocity (m/s)	0.46 ± 0.22	0.54 ± 0.22	0.59 ± 0.18 *	0.51 ± 0.22	0.69 ± 0.17 †*	0.83 ± 0.25 †*^	1.39 ± 0.06
Cadence (steps/min)	80.9 ± 16.5	86.0 ± 18.7	87.0 ± 16.1	83.9 ± 17.2	95.80 ± 11.8 †*	105.0 ± 12.1 †*^	113.8 ± 4.3
Step length – operated limb (m)	0.33 ± 0.12	0.36 ± 0.12	0.39 ± 0.10	0.33 ± 0.11	0.39 ± 0.11 *	0.42 ± 0.10 *	0.73 ± 0.02
Gait cycle length (m)	0.73 ± 0.22	0.79 ± 0.20	0.84 ± 0.19 *	0.74 ± 0.23	0.89 ± 0.16 *	0.96 ± 0.21 †*^	1.47 ± 0.08
Step width (m)	0.19 ± 0.04	0.18 ± 0.03	0.18 ± 0.04	0.18 ± 0.04	0.17 ± 0.04	0.18 ± 0.04	0.13 ± 0.01
Pain (VAS score)	6.70 ± 2.60	4.90 ± 2.40	3.50 ± 2.50 *	7.70 ± 2.30	2.90 ± 1.60 †*	1.50 ± 1.40 †*	–

Data are means ±SD. p <0.05 denotes statistically significant differences, † study patients vs. corresponding feature in control patients, * after vs. before surgery in a group of patients, ^ 6 months vs. 1 month after surgery in a group

aStandard rehabilitation, bproprioceptive neuromuscular facilitation (PNF) physiotherapy, cresults representing normal values from healthy adults (database of BTS Bioengineering Corp., Brooklyn, NY), VAS, visual analog scale

A p-value < 0.05 defined statistically significant differences. The effect size was calculated based on Pearson's (r) and Cohen's (d) coefficients.

3   Results

3.1   Stance and Swing Phase Duration

The stance phases in both study and control patients, concerning both operated and non-operated limbs, were comparable before surgery and were much, on average, 50 s longer than the reference in the age-matched healthy subjects. Both 1 and 6 months after surgery in the study patients, stance phase of the operated limb was significantly shorter (0.80 ± 0.14 s and 0.72 ± 0.11 s, respectively) than those in the control patients after surgery (1.08 ± 0.62 s and 1.03 ± 0.50 s, respectively). There was a large effect size in the study patients ($r = 0.44$ and $r = 0.62$) after both 1 and 6 months post-surgery. The stance phase also was shorter in these patients 6 months after surgery compared with its duration 1 month after surgery, the effect not observed in the control patients. Further, similar changes in stance phase were also present in the non-operated limb (Table 1).

Before surgery, the relation of stance phase to swing phase in the patients examined was distorted, with the stance constituting about 68% of the gait cycle. The contribution of the stance to the gait cycle decreased significantly to 63% after 1 month and 62% 6 months after surgery in the study patients, whereas this contribution failed to change appreciably in the control patients. The relation between the stance and swing phases in the study patients was within the norm for a healthy 65+ years of age adult 6 months after surgery. There was a large effect size in these patients ($d = 1.04$ and $r = 0.6$) after both 1 and 6 months post-surgery. In the control patients, by contrast, relation between stance and swing phases was distorted to the same extent before and 1 and 6 months after surgery (Table 1).

There were no major changes in swing phase duration before and after surgery in either control or study patients.

3.2   Gait Cycle Duration and Double-Limb Support Phase

Gait cycle duration was longer before surgery in both control and study patients than the reference in healthy subjects. It shortened significantly in the study, but not control, patients after surgery; the shortening progressed with time after surgery amounting to about 22% 1 month and 32% 6 months after surgery ($p < 0.05$). There also was a large effect size ($r = 34, r = 0.55$).

Likewise, double-limb support phase, which was in a range of 19–20 s before surgery in both groups of patients, shortened significantly and progressively 1 and 6 months after surgery by about 33% and 43% in the study, but not control, patients ($p < 0.05$), with a large effect size ($d = 0.9$ and $r = 0.6$) (Table 1).

3.3   Swing Phase Velocity, Mean Gait Velocity, and Cadence

Before surgery, swing phase velocity was smaller in both control and study patients than the reference in healthy subjects. Although it increased 6 months after surgery in the control patients, the increase was distinctly greater, by 22% and 42% 1 and 6 months, respectively, and significantly progressive ($p < 0.05$) with a large effect size ($d = 0.8$) 6 months after surgery in the study patients (Table 1).

Mean gait velocity amounted to about 0.50 m/s in both control and study patients before surgery, which was about threefold smaller than the reference in healthy subjects. Akin to swing velocity, gait velocity increased 6 months after surgery in control patients, and the increase was distinctly greater, by 35% and 63% 1 and 6 months, respectively, and significantly progressive ($p < 0.05$) with a large effect size ($d = 0.75$,

$d = 1.08$) 6 months after surgery in the study patients, nearing the gait velocity of the reference healthy subjects.

Likewise, cadence, expressed in steps *per* minute, was, on average, comparable – 81 steps/min in control and 84 steps/min in study patients – but smaller compared with the reference of about 114 steps/min in healthy subjects. While the number of steps failed to change appreciably in the control subjects, it significantly and progressively increased by 14% and 25% 1 and 6 months after surgery in the study patients ($p < 0.05$), with a large effect size ($r = 0.3$, $d = 1.3$) (Table 1).

3.4   Step Length, Gait Cycle Length, and Step Width

Step and gait cycle lengths in patients qualified for surgery were in a range of 0.33 m and 0.73 m, respectively, both about twofold shorter than the respective reference in healthy subjects. Both variables tended to increase in control patients 6 months after surgery; the increases were distinctly and progressively greater than 1 and 6 months after surgery in the study patients ($p < 0.05$), although they remained short of the reference levels in healthy subjects (Table 1).

Step width in both control and study patients suffering from knee osteoarthritis was greater than the reference in healthy subjects, on average, 20 cm vs. 13 cm, respectively. There were no appreciable changes in step width in either group of patients after surgery.

Individual-targeted neuromuscular facilitation treatment leads to a significant decrease in pain perception score, which was already evident 1 month after knee arthroplasty surgery. A decrease in pain symptoms was sustained over time as pain further tapered off 6 months after surgery (Table 1).

4   Discussion

The influence on gait pattern of knee joint replacement has already been investigated. However, the effectiveness of different rehabilitation

protocols is largely unknown. This study demonstrates that the individual-targeted physiotherapy program introduced in the first month after surgery significantly improves gait pattern of a patient who underwent a knee joint replacement. The technique of proprioceptive neuromuscular facilitation we employed in the study is in accord with the rehabilitation guidelines for knee arthroplasty of the Osteoarthritis Research Society International (OARSI) and Ottawa Panel evidence-based clinical practice (Peter et al. 2011; Zhang et al. 2010; Brosseau et al. 2005).

Artz et al. (2013) have investigated the effects of various routines of physiotherapy in patients discharged after total knee and hip arthroplasty from 24 high-volume NHS orthopedic clinics in England and Wales. There were eleven clinics in which the majority of patients participated in group exercises and five in which individual physiotherapy was employed. The authors conclude that although none of the clinics refer postoperative patients for rehabilitation as a standard procedure, physiotherapy after knee arthroplasty is a more common practice than after hip arthroplasty and is more commonly executed in group exercises. Oatis et al. (2014) have carried out a 6-month-long postoperative observation of knee arthroplasty patients in different hospitals across the USA. It turns out that there is a significant difference concerning the type of intervention used, number of repetitions, resistance, frequency of exercises, and the time of rehabilitation commencement. A meta-analysis concerning the effectiveness of physiotherapy in postoperative knee arthroplasty has demonstrated the lack of improvement in gait pattern in patients performing general exercises. However, gait velocity is improved in patients practicing weight bearing on the operated limb and various ways of walking (Artz et al. 2015). Hausdorf and Kang have demonstrated that gait cycle duration in the elderly associates with the strength of quadriceps, the range of movability, and the gait velocity (Hausdorff et al. 2011; Kang and Dingwell 2008). Patients experiencing knee joint pain tend to walk much slower than healthy persons do, which enhances the risk of losing stability while walking. Those and other studies demonstrate that gait velocity, step length, and range of limb motion are persistently reduced post-surgery, and the slow developing improvements fall short of motion level seen in healthy subjects (Casartelli et al. 2013; Alnahdi et al. 2011; Bennett et al. 2008).

In the present study, we demonstrate that an individually tailored physiotherapy program, based on proprioceptive neuromuscular facilitation, significantly improved gait velocity by shortening the stance and double support phases. Gait velocity improved already 1 month after surgery, and the improvement was sustained and even advanced 6 months after surgery in patients subjected to individually tailored physiotherapy, but not in patients subjected to standard physiotherapy. Gait velocity of healthy adults between 30 and 60 years of age varies between 1.2 and 1.5 m/s (Blanke and Hageman 1989). It decreases to 1.0–1.2 m/s in healthy elderly (Brach et al. 2007). Newman et al. (2003) have demonstrated that gait velocity required for safe street crossing is 1.2 m/s. In a study by Alice et al. (2015), gait velocity failed to improve in a half of the 90 patients 3 months after knee joint arthroplasty, and it was even worse after than before surgery in every fifth patient. Those patients, however, were not subject to an individual physiotherapy. Likewise, Hiyama et al. (2015) have demonstrated that surgery alone may not be enough for a clear benefit in gait pattern. Taş et al. (2014) have demonstrated the following changes in gait kinematics in severe knee osteoarthritis patients in stage 3: reductions in gait velocity, cadence, step length, and gait cycle length and increases in the duration of gait cycle, step, one-legged stance phase, double support phase, and stance phase. Similar observations concerning the extension of stance and double support phases in severe knee arthritis, compared with healthy, age-matched persons, have been made by other authors (Harding et al. 2012; Kiss 2011; Kiliçoğlu et al. 2010; Astephen et al. 2008). The extension of double support may constitute a compensatory mechanism aiming at supporting body weight by both lower limbs and thus reducing the affected knee joint loading while walking. It also may lead to an increase in both stance

phase and gait velocity (Taş et al. 2014). According to Hamacher et al. (2011), patients who fall also feature extended stance phase, decreased gait velocity, and shortened step length. Therefore, rehabilitation program after knee arthroplasty should combine interventions aimed at improving stability, proprioception, neuromuscular control, and facilitation of different gait phases. A change of spatiotemporal gait pattern may influence walking stability and ameliorate balance disorders.

Stan and Orban (2014) have demonstrated a significant difference in the average duration of loading between operated and non-operated knee joints, being 1.03 s and 1.08 s, respectively. In healthy persons, the duration of 0.8 s was distinctly lower, compared to that in patients both before and after surgery. The extension of stance duration of non-operated limb together with decreased gait velocity seems a postoperative strategy for load avoidance. Such asymmetry may, however, lead to overloading of the other operated limb. In the present study, duration of loading of the operated limb shortened from 1.03 ± 0.34 s to 0.72 ± 0.11 s 6 months after surgery, and the discrepancy between operated and non-operated limb was leveled off in patients subjected to individual neuromuscular facilitation, but not standard, physiotherapy. The findings underscore the potential benefit to be gained from individually tailored therapy.

McClelland et al. (2011) have actually observed a worsening of gait velocity, cadence, and step length in patients who were 12 months after knee replacement surgery, compared with healthy persons. That observation contrasts sharply with our present findings demonstrating an overall improvement in gait kinematics after knee arthroplasty, although the follow-up in the present study did not extend beyond 6 months. Noteworthy, the distinguishing factor between these studies was the lack of individual neuromuscular facilitation physiotherapy in the former and its use in the latter study. Neuromuscular facilitation physiotherapy should, however, involve a therapist as according to Szöts et al. (2015), as many as 70% of patients admitted for elective knee joint arthroplasty do not regularly perform recommended exercises at home in the first months after surgery. One of the advantages of such is a significant reduction in pain perception during the time following knee arthroplasty, which we noted in this study and which should not be underestimated. Post-surgery pain catastrophizing is a psychological construct that leads to preoccupation with fear of pain and a related avoidance of mobility and recommended exercises, all of which may worsen surgery outcomes (Burns et al. 2015).

In conclusion, we believe we have shown in this study that individually tailored physiotherapy program consisting of proprioceptive neuromuscular facilitation is effectively advantageous in improving gait kinematic pattern as well as psychological aspects related to pain perception in patients after knee arthroplasty surgery. Surgical correction alone, without proper rehabilitation accompanied by therapist help, is insufficient to obtain optimum outcomes.

Conflicts of Interest The authors declare no conflicts of interest in relation to this article.

References

Alice B-M, Stéphane A, Yoshisama SJ, Pierre H, Domizio S, Hermes M, Katia T (2015) Evolution of knee kinematics three months after total knee replacement. Gait Posture 41(2):624–629

Alnahdi AH, Zeni JA, Snyder-Mackler L (2011) Gait after unilateral total knee arthroplasty: frontal plane analysis. J Orthop Res 29(5):647–652

Artz N, Dixon S, Wylde V, Beswick A, Blom A, Gooberman-Hill R (2013) Physiotherapy provision following discharge after total hip and total knee replacement: a survey of current practice at high-volume NHS hospitals in England and Wales. Musculoskeletal Care 11(1):31–38

Artz N, Elvers KT, Lowe CM, Sackley C, Jepson P, Beswick AD (2015) Effectiveness of physiotherapy exercise following total knee replacement: systematic review and meta-analysis. BMC Musculoskelet Disord 16:15

Astephen JL, Deluzio KJ, Caldwell GE, Dunbar MJ (2008) Biomechanical changes at the hip, knee, and ankle joints during gait are associated with knee osteoarthritis severity. J Orthop Res 26(3):332–341

Bennett D, Humphreys L, O'Brien S, Kelly C, Orr JF, Beverland DE (2008) Gait kinematics of age-stratified hip replacement patients – a large scale, long-term follow-up study. Gait Posture 28(2):194–200

Beswick AD, Wylde V, Gooberman-Hill R, Blom A, Dieppe P (2012) What proportion of patients report long-term pain after total hip or knee replacement for osteoarthritis? A systematic review of prospective studies in unselected patients. BMJ Open 2(1):e000435

Blanke DJ, Hageman PA (1989) Comparison of gait of young men and elderly men. Phys Ther 69(2):144–148

Brach JS, Studenski SA, Perera S, VanSwearingen JM, Newman AB (2007) Gait variability and the risk of incident mobility disability in community-dwelling older adults. J Gerontol A Biol Sci Med Sci 62 (9):983–988

Brosseau L, Wells GA, Tugwell P et al (2005) Ottawa panel evidence-based clinical practice guidelines for therapeutic exercises and manual therapy in the management of osteoarthritis. Phys Ther 85(9):907–971

Burns LC, Ritvo SE, Ferguson MK, Clarke H, Seltzer Z, Katz J (2015) Pain catastrophizing as a risk factor for chronic pain after total knee arthroplasty: a systematic review. J Pain Res 8:21–32

Casartelli NC, Item-Glatthorn JF, Bizzini M, Leunig M, Maffiuletti NA (2013) Differences in gait characteristics between total hip, knee, and ankle arthroplasty patients: a six-month postoperative comparison. BMC Musculoskelet Disord 14:176

Hamacher D, Singh NB, Van Dieën JH, Heller MO, Taylor WR (2011) Kinematic measures for assessing gait stability in elderly individuals: a systematic review. J R Soc Interface 8(65):1682–1698

Harding GT, Hubley-Kozey CL, Dunbar MJ, Stanish WD, Astephen Wilson JL (2012) Body mass index affects knee joint mechanics during gait differently with and without moderate knee osteoarthritis. Osteoarthr Cartil 20(11):1234–1242

Hausdorff JM, Rios DA, Edelberg HK (2011) Gait variability and fall risk in community-living older adults: a 1-year prospective study. Arch Phys Med Rehabil 82(8):1050–1056

Hiyama Y, Asai T, Wada O, Maruno H, Nitta S, Mizuno K, Iwasaki Y, Okada S (2015) Gait variability before surgery and at discharge in patients who undergo total knee arthroplasty: a cohort study. PLoS One 10(1):e0117683

Kang HG, Dingwell JB (2008) Separating the effects of age and walking speed on gait variability. Gait Posture 27(4):572–577

Kiliçoğlu O, Dönmez A, Karagülle Z, Erdoğan N, Akalan E, Temelli Y (2010) Effect of balneotherapy on temporospatial gait characteristics of patients with osteoarthritis of the knee. Rheumatol Int 30 (6):739–747

Kiss RM (2011) Effect of severity of knee osteoarthritis on the variability of gait parameters. J Electromyogr Kinesiol 21(5):695–703

Lee A, Park J, Lee S (2015) Gait analysis of elderly women after total knee arthroplasty. J Phys Ther Sci 27(3):591–595

McClelland JA, Webster KE, Feller JA, Menz HB (2011) Knee kinematics during walking at different speeds in people who have undergone total knee replacement. Knee 18(3):151–155

Newman AB, Haggerty CL, Kritchevsky SB, Nevitt MC, Simonsick EM, Health ABC, Collaborative Research Group (2003) Walking performance and cardiovascular response: associations with age and morbidity – the health, aging and body composition study. J Gerontol A Biol Sci Med Sci 58(8):715–720

Oatis CA, Li W, DiRusso JM, Hoover MJ, Johnston KK, Butz MK, Phillips AL, Nanovic KM, Cummings EC, Rosal MC, Ayers DC, Franklin PD (2014) Variations in delivery and exercise content of physical therapy rehabilitation following total knee replacement surgery: a cross-sectional observation study. Int J Phys Med Rehabil 5(Suppl 5):pii: 002

Peter WF, Jansen MJ, Hurkmans EJ, Bloo H, Dekker J, Dilling RG, Hilberdink W, Kersten-Smit C, de Rooij M, Veenhof C, Vermeulen HM, de Vos RJ, Schoones JW, Vliet Vlieland TP, Guideline Steering Committee – Hip and Knee Osteoarthritis (2011) Physiotherapy in hip and knee osteoarthritis: development of a practice guideline concerning initial assessment,

treatment and evaluation. Acta Reumatol Port 36 (3):268–281

Stan G, Orban H (2014) Human gait and postural control after unilateral total knee arthroplasty. Mædica (Buchar) 9(4):356–360

Szöts K, Pedersen PU, Hørdam B, Thomsen T, Konradsen H (2015) Physical health problems experienced in the early postoperative recovery period following total knee replacement. Int J Orthop Trauma Nurs 19 (1):36–44

Taş S, Güneri S, Baki A, Yıldırım T, Kaymak B, Erden Z (2014) Effects of severity of osteoarthritis on the temporospatial gait parameters in patients with knee osteoarthritis. Acta Orthop Traumatol Turc 48 (6):635–641

Zhang W, Nuki G, Moskowitz RW, Abramson S, Altman RD, Arden NK, Bierma-Zeinstra S, Brandt KD, Croft P, Doherty M, Dougados M, Hochberg M, Hunter DJ, Kwoh K, Lohmander LS, Tugwell P (2010) OARSI recommendations for the management of hip and knee osteoarthritis: part III: changes in evidence following systematic cumulative update of research published through January 2009. Osteoarthr Cartil 18(4):476–499

Adv Exp Med Biol - Clinical and Experimental Biomedicine (2018) 1: 11–17
https://doi.org/10.1007/5584_2018_193
© Springer International Publishing AG, part of Springer Nature 2018
Published online: 6 April 2018

Does Patient-Specific Instrumentation Improve Femoral and Tibial Component Alignment in Total Knee Arthroplasty? A Prospective Randomized Study

Artur Stolarczyk, Lukasz Nagraba, Tomasz Mitek, Magda Stolarczyk, Jarosław Michał Deszczyński, and Maciej Jakucinski

Abstract

Alignment of the prosthesis is one of the most significant factors that affect the long-term clinical outcome following total knee arthroplasty (TKA). There is conflicting evidence whether patient-specific instrumentation (PSI) for TKA improves the component position compared to standard instrumentation. This study aimed to compare the rotational alignment of the femoral and tibial components in TKA patients when performed with either conventional or PSI. Sixty patients with primary knee osteoarthritis were randomly divided into two groups treated surgically with TKA: one with conventional instrumentation and the other with the Visionaire PSI system (Smith and Nephew, Memphis, TN). Computerized tomography (CT) and X-ray imaging were performed pre- operatively and 12 weeks after surgery. The rotational alignment of the femoral and tibial component in all patients was assessed post-surgically using CT imaging according to the Berger protocol. Both groups were clinically assessed in a blinded fashion using the Knee Society Score (KSS) and a visual analog scale (VAS). Fifty-eight patients were prospectively assessed. The mean postsurgical follow-up was 3.0 ± 0.4 months. CT images did not reveal any significant improvement in the rotational alignment of the implant components between the groups. X-rays revealed a significant improvement in the deviation from the optimal alignment range of the femoral component in the coronal plane in both groups. Patients operated with Visionaire PSI assistance had poorer functional outcomes. We conclude that there were no improvements in clinical outcomes or knee component

A. Stolarczyk (✉) and J. M. Deszczyński
Department of Clinical Rehabilitation, Warsaw Medical University, Warsaw, Poland
e-mail: artur.stolarczyk@wum.edu.pl

L. Nagraba, T. Mitek, and M. Stolarczyk
Department of Orthopedics and Rehabilitation, Warsaw Medical University, Warsaw, Poland

M. Jakucinski
Department of Radiology, Warsaw Medical University, Warsaw, Poland

alignment in patients treated with PSI compared with those treated with standard instruments. In addition, clinical and functional assessment showed inferior results in terms of KSS and VAS scores at the midterm follow-up in patients treated with PSI.

Keywords

Computerized tomography · Femoral component rotation · Knee arthroplasty · Patient-specific instrumentation · Rotational alignment · Tibia

1   Introduction

The primary objective of total knee arthroplasty (TKA) is to improve knee function and relieve pain. Another important factor is the implant longevity that determines the long-term effect of surgery (Mahaluxmivala et al. 2001; Ritter et al. 1994; Jeffery et al. 1991; Petersen and Engh 1988; Rand and Coventry 1988; Lotke and Ecker 1977). A rotational implant alignment is crucial for a long-term success of TKA. It has been reported that the vast majority of TKA procedures have internal rotational errors in the femoral/tibial component alignment (Nicoll and Rowley 2010). A frequent presence of such errors has also been evidenced in a study in which all TKA cases demonstrated knee stiffness due to tibial component misalignment consisting of excessive internal rotation (Bedard et al. 2011).

Patient-specific instrumentation (PSI) for TKA has emerged as a more precise alternative to standard instrumentation. The PSI, employing preoperative three-dimensional imaging, provides personalized jigs based on the anatomic landmarks to determine the placement of conventional cutting blocks, which helps maintain the correct positional relationship of components. Unlike the PSI, standard instrumentation relies on intramedullary or extramedullary alignment rods and the surgeon's judgment for placement of conventional cutting blocks. There is scarce and conflicting evidence in the literature that compares the accuracy of PSI to standard

instrumentation (Stronach et al. 2014; Conteduca et al. 2013; Conteduca et al. 2012; Ng et al. 2012; Nunley et al. 2012). The existing studies emphasize a coronal alignment and largely rely on the long-limb radiographs to establish the measurements. However, to evaluate the true accuracy of PSI, a preoperative 3D plan should be directly compared with the postoperative 3D prosthetic alignment.

The present study was designed to compare the degree of rotation of the femoral and tibial components in TKA patients who underwent joint replacement with either conventional or PSI instrumentation. We also assessed the midterm clinical results.

2   Methods

2.1   Patients and Intervention

The Ethics Committee of Warsaw Medical University in Warsaw, Poland, approved the study. All patients gave informed consent to participate in the study. The study included 60 patients who had been admitted to the Department of Orthopedics and Rehabilitation of Warsaw Medical University between November 2012 and December 2014. The patients' mean age was 69.9 ± 6.5 (SD) years, and they were all qualified for the TKA procedure due to primary knee osteoarthritis. The patients were randomly assigned to one of the two groups. The study group underwent TKA with the Visionaire PSI (VISIONAIRE; Smith and Nephew, Memphis, TN), and the second control group underwent TKA using a conventional instrumentation. A block randomization with a block size of six was performed using a computer-generated random number list that was prepared by an investigator who had no clinical involvement in the trial. The sequence was concealed until all data were analyzed. Each patient in the PSI and conventional groups had CT and X-ray scans on the operated knee, 3 weeks after surgery.

The methodology of the study precluded the possibility of a blinded patient-surgeon relationship. Therefore, to reduce the risk of a systematic bias related to the analysis of outcomes, the radiologist who assessed the degree of rotation in the femoral and tibial components was blinded concerning the technique used in a given patient.

2.2   Outcomes

The primary outcome was the rotational alignment of the femoral and tibial component, based on using the Berger protocol (Berger and Crossett 1998) and assessed postoperatively with CT imaging scans. The findings in the PSI and conventional groups were compared. Secondary endpoints focused on the clinical assessment, including function, which was evaluated in both groups with the Knee Society Score (KSS), and perception of pain evaluated on a visual analog score (VAS). Tertiary endpoints included the assessment of surgery duration, the length of the surgical scar, and the duration of postoperative hospital stay.

2.3   Radiologic Evaluation

Three independent radiologists were involved in the analysis of the 60 sets of CT scans in a random order and in separate rooms to prevent a possible recollection of the previous viewing. Each radiologist was asked to evaluate CT scans according to the Berger protocol. The readings were repeated 6 weeks later to assess intra-observer variability. The radiologists were not provided with any feedback concerning their assessments, and the CT scans were unavailable to them between the readings. A comparison of the imaging results is given in Table 2.

2.4   Statistical Evaluation

A sample size of 30 patients in each group was chosen based upon a recently published study that showed that a computer navigation had some significant benefits when compared with standard instrumentation (Woolson et al. 2014). The evaluation of continuous data was performed with a t-test and the Mann-Whitney U test for normally and non-normally distributed variables, respectively. The χ^2 or Fisher's exact test were used for the analysis of dichotomous outcomes, as appropriate. The relative risk (RR) or mean difference (MD) with 95% confidence intervals (95% CI) was calculated using StatsDirect v2.7.8b software. Differences between the groups were considered statistically significant when a p-value was <0.05, the 95%CI for RR did not exceed 1.0, or the RR for MD did not exceed 0. The results were analyzed using the intention to treat analysis (ITT) for postoperative data and the available case analysis for clinical data after a 12-week long follow-up. Computer software "R" v2.13.1 was used for all analyses.

3   Results

3.1   Study Flowchart

The study included 60 patients who, according to the randomization list, were assigned to the study (TKA with PSI) or control group, 30 patients each. All the data required for the assessment of primary and tertiary endpoints were successfully obtained from all the participants (ITT analysis). With regard to the secondary endpoints, data were obtained from 58 patients (two patients failed to report for follow-up). Therefore, the available case analysis was carried out (Fig. 1). Table 1 lists the patients' demographic data, the main KSS assessment variables, and the subjective pain scale prior to surgery in the study group and in the control group (conventional TKA).

3.2   Radiologic and CT Evaluations

The CT scans were analyzed to assess the rotation of both femoral and tibial components. In the case of the femoral component, a slightly better positioning was observed in patients who had

Table 1 Patients' demographics prior to total knee arthroplasty surgery

	Study group (n = 30)	Control group (n = 30)	p-value
Age (ys)	70.2 ± 5.9	69.6 ± 7.1	NS
Female/male	22/8	18/12	NS
BMI (kg/m^2)	30.4 ± 4.4	31.6 ± 5.4	NS
KSS – knee	34.1 ± 11.8	34.2 ± 14.5	NS
KSS – function	48.0 ± 19.8	49.7 ± 18.2	NS
VAS (pain 1–10)	7.9 ± 1.5	7.9 ± 1.8	NS

Study group – patient-specific instrumentation (PSI) and control group – conventional instrumentation
Data are means ±SD. *BMI* body mass index, *KSS* Knee Society Score, *VAS* visual analog scale, *NS* nonsignificant

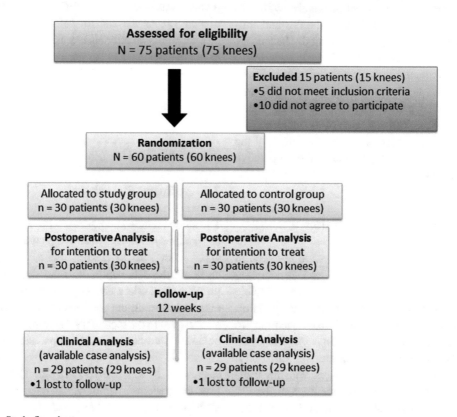

Fig. 1 Study flowchart

undergone TKA with PSI ($0.2° ± 5.4°$ vs. $-0.9° ± 6.5°$; MD = 0.2°, 95% CI -1.27 - 1.67°). No statistically significant differences were found between the study and control groups in terms of mechanical lateral distal femur angle (mLDFA), medial proximal tibia angle (MPTA), or rotational alignment of tibial component. All these data are presented in Table 2.

3.3 Clinical Assessment: Secondary Endpoints

Clinical assessments performed, on average, in the 12th week postoperative follow-up did not reveal any significant differences between the two groups. The results of the assessments of KSS parameters and subjective pain were also

Table 2 Comparison of imaging results between the study group – patient-specific instrumentation (PSI) and the control group – conventional instrumentation

Variable	Study group (n = 29)	Control group (n = 29)	MD (95% CI)
Femorotibial mechanical axis	1.0 ± 3.1°	0.8 ± 2.7°	0.20 (−1.3, 1.7)
mLDFA	89.5 ± 1.5°	89.5 ± 2.5°	0.00 (−1.06, 1.06)
Rotational alignment of femoral component	0.2 ± 5.4°	−0.9 ± 6.5°	1.10 (−1.98, 4.18)
MPTA	88.9 ± 2.1°	89.1 ± 1.7°	−0.20 (−1.18, 0.78)
Rotational alignment of tibial component	−0.6 ± 5.0°	−1.6 ± 6.1°	1.00 (−1.87, 3.87)
Slope of tibial plateau	−4.6 ± 2.3°	−4.0 ± 2.6°	−0.6 (−1.86, 0.66)

Data are means ±SD. *MD* mean difference with 95% confidence intervals, *mLDFA* mechanical lateral distal femur angle, *MPTA* medial proximal tibia angle

Table 3 Comparison of clinical outcomes, based on scale scores, between the study group – patient-specific instrumentation (PSI) and the control group – conventional instrumentation

	Study group (n = 29)	Control group (n = 29)	MD (95% CI)
KSS – knee	74.1 ± 20.4	69.5 ± 16.9	4.60 (−5.04–14.24)
KSS – function	69.3 ± 15.8	68.2 ± 16.3	1.10 (−7.16–9.36)
VAS [pain 1–10]	2.2 ± 1.7	2.0 ± 1.5	0.20 (−0.63–1.03)

Data are means ±SD. *MD* mean difference with 95% confidence intervals, *KSS* Knee Society Score, and *VAS* visual analog scale

comparable (Table 3). At the follow-up, poorer clinical outcomes, according to KSS and VAS scores, were observed in the patients who underwent TKA with PSI.

3.4 Evaluation of Perioperative and Intraoperative Parameters: Tertiary Endpoints

Tertiary endpoints were directly related to perioperative and intraoperative parameters, and they revealed certain significant differences between the two groups. The PSI-TKA group was found to feature, among other factors, a prolonged duration of surgery by more than 30 min (MD 32 min; 95%CI 24.7–39.3 min), a difference in the length of the surgical incision that was more than 2 cm longer (MD 2.1 cm; 95%CI: 1.0–3.2 cm), and approximately 2 additional days of hospital stay (MD 2.2 days; 95%CI: 0.4–4.0 days) (Table 4).

4 Discussion

The application of PSI is hypothesized to help achieve the optimal rotational placement of tibial and femoral components in TKA procedures. This prospective, randomized trial demonstrates that Visionaire PSI in patients undergoing TKA had no influence on the femoral and tibial axial rotation as assessed by CT imaging. Further, the choice of the surgical technique had no bearing upon clinical function or pain assessment after a 12-week follow-up.

According to the study by Heyse and Tibesku (2015), application of PSI for TKA considerably reduces the percentage of deviations from the optimal rotational placement of the tibial component in the magnetic resonance imaging assessments. Those authors have emphasized that the anatomy of the proximal end of the tibia is highly variable across the population, a feature that hinders the selection of the optimal points of reference and PSI production. In another article, the same authors have demonstrated a significant improvement in the rotational positioning of the femoral component in the PSI group compared to patients undergoing conventional surgery: akin to the tibia, rotation of the femoral component was assessed by magnetic resonance imaging (Heyse and Tibesku 2014). The incompatibility of the results of those studies with the current study may be influenced by the operator experience and the level of the operator's comfort with each surgical technique. On the other hand,

Table 4 Comparison of perioperative and intraoperative parameters between the study group – patient-specific instrumentation (PSI) and the control group – conventional instrumentation

	Study group (n = 30)	Control group (n = 30)	MD (95% CI)
Duration of surgery (min)	98 ± 15	66 ± 14	32 (24.66–39.34)
Length of incision (cm)	21.4 ± 2.1	19.3 ± 2.4	2.10 (0.96–3.24)
Duration of hospital stay (days)	14.2 ± 4.7	12.0 ± 2.1	2.20 (0.36–4.04)

Data are means ±SD. *MD* mean difference with 95% confidence intervals

since the patients of those previous studies did not undergo clinical observation over time, no definite conclusions about the improvement of function can be reached.

The achievement of optimal alignment in the coronal, sagittal, and rotational plane is closely related to the performance of correct bone incisions in the tibia and femur adapted individually to each patient (Jenny et al. 2005; Laskin 2003; Reed and Gollish 1997; Jeffery et al. 1991). As an innovative concept capitalizing on the various advantages of computer technologies, PSI makes it possible to transfer some stages of surgery to preoperative planning. Compared to computer navigation, for instance, PSI does not entail significant additional costs and shortens the duration of surgery. In the current study, duration of surgery was shorter by 30 min in the control group. Likewise, the entire hospital stay of patients in the control group was 2 days shorter compared to the PSI-TKA group.

Despite numerous hypothetical and technical advantages related to the use of PSI, evidence-based data on the benefit of its application in terms of the patients' improved clinical function are inconsistent. Some authors have concluded that there are such benefits, while others failed to demonstrate any differences compared to classical instrumentation (Fu et al. 2015). The current study also failed to confirm any advantageous effect of this type of instrumentation on the improvement of clinical function in patients after a 12-week postoperative follow-up.

4.1 Strengths and Limitations

We used the acceptable methods to generate the allocation sequence and allocation concealment. We then strived to maintain the blinding of data management and data analyses throughout the study. In addition, the length of follow-up was appropriate. Data on radiological outcomes were obtained for all patients and data on clinical outcomes for more than the 93% of them. All of these features minimize the risk of systematic bias.

We believe that this is the largest randomized trial to date that have assessed the efficacy of PSI by means of CT imaging. The application of CT together with a clinical assessment made it possible to evaluate and compare the efficacy of the two surgical techniques and could also suggest the mechanisms underlying the differences between the two groups.

In conclusion, the results of this study found that PSI and standard surgical techniques were comparable in terms of component alignment and patient-reported outcomes, although operative time and length of stay were longer in the PSI-TKA group. A number of studies have failed to demonstrate any advantages resulting from the use of PSI. Our current findings also failed to reveal any positive effects of PSI in terms of improved rotational component alignment or improved clinical outcomes. In light of the inconsistent data in the literature, further studies are still needed to resolve the contentious issue.

Conflicts of Interest The authors declared no conflicts of interest in relation to this article.

References

Bedard M, Vince KG, Redfern J, Collen SR (2011) Internal rotation of the tibial component is frequent in stiff total knee arthroplasty. Clin Orthop Relat Res 469 (8):2346–2355

Berger RA, Crossett LS (1998) Determining the rotation of the femoral and tibial components in total knee arthroplasty: a computed tomography technique. Oper Tech Orthop 8:128–133

Conteduca F, Iorio R, Mazza D, Caperna L, Bolle G, Argento G, Ferretti A (2012) Are MRI-based, patient matched cutting jigs as accurate as the tibial guides? Int Orthop 36(8):1589–1593

Conteduca F, Iorio R, Mazza D, Caperna L, Bolle G, Argento G, Ferretti A (2013) Evaluation of the accuracy of a patient-specific instrumentation by navigation. Knee Surg Sports Traumatol Arthrosc 21(10):2194–2199

Fu H, Wang J, Zhou S, Cheng T, Zhang W, Wang Q, Zhang X (2015) No difference in mechanical alignment and femoral component placement between patient-specific instrumentation and conventional instrumentation in TKA. Knee Surg Sports Traumatol Arthrosc 23(11):3288–3295

Heyse TJ, Tibesku CO (2014) Improved femoral component rotation in TKA using patient-specific instrumentation. Knee 21(1):268–271

Heyse TJ, Tibesku CO (2015) Improved tibial component rotation in TKA using patient-specific instrumentation. Arch Orthop Trauma Surg 135(5):697–701

Jeffery RS, Morris RW, Denham RA (1991) Coronal alignment after total knee replacement. J Bone Joint Surgery Br 73(5):709–714

Jenny JY, Clemens U, Kohler S, Kiefer H, Konermann W, Miehlke RK (2005) Consistency of implantation of a total knee arthroplasty with a non-image-based navigation system: a case-control study of 235 cases compared with 235 conventionally implanted prostheses. J Arthroplast 20(7):832–839

Laskin RS (2003) Instrumentation pitfalls: you just can't go on autopilot! J Arthroplast 18(3 Suppl 1):18–22

Lotke PA, Ecker ML (1977) Influence of positioning of prosthesis in total knee replacement. J Bone Joint Surg Am 59(1):77–79

Mahaluxmivala J, Bankes MJ, Nicolai P, Aldam CH, Allen PW (2001) The effect of surgeon experience on component positioning in 673 press fit condylar posterior cruciate-sacrificing total knee arthroplasties. J Arthroplast 16(5):635–640

Ng VY, DeClaire JH, Berend KR, Gulick BC, Lombardi AV Jr (2012) Improved accuracy of alignment with patient-specific positioning guides compared with manual instrumentation in TKA. Clin Orthop Relat Res 470(1):99–107

Nicoll D, Rowley DI (2010) Internal rotational error of the tibial component is a major cause of pain after total knee replacement. J Bone Joint Surg Br 92(9):1238–1244

Nunley RM, Ellison BS, Zhu J, Ruh EL, Howell SM, Barrack RL (2012) Do patient-specific guides improve coronal alignment in total knee arthroplasty? Clin Orthop Relat Res 470(3):895–902

Petersen TL, Engh GA (1988) Radiographic assessment of knee alignment after total knee arthroplasty. J Arthroplast 3(1):67–72

Rand JA, Coventry MB (1988) Ten-year evaluation of geometric total knee arthroplasty. Clin Orthop Relat Res (232):168–173

Reed SC, Gollish J (1997) The accuracy of femoral intramedullary guides in total knee arthroplasty. J Arthroplast 12(6):677–682

Ritter MA, Faris PM, Keating EM, Meding JB (1994) Postoperative alignment of total knee replacement. Its effect on survival. Clin Orthop Relat Res (299):153–156

Stronach BM, Pelt CE, Erickson JA, Peters CL (2014) Patient-specific instrumentation in total knee arthroplasty provides no improvement in component alignment. J Arthroplast 29(9):1705–1708

Woolson ST, Harris AH, Wagner DW, Giori NJ (2014) Component alignment during total knee arthroplasty with use of standard or custom instrumentation: a randomized clinical trial using computed tomography for postoperative alignment measurement. J Bone Joint Surg Am 96(5):366–372

Adv Exp Med Biol - Clinical and Experimental Biomedicine (2018) 1: 19–29
https://doi.org/10.1007/5584_2018_186
© Springer International Publishing AG, part of Springer Nature 2018
Published online: 4 April 2018

Preoperative Rehabilitation in Lung Cancer Patients: Yoga Approach

Giovanni Barassi, Rosa Grazia Bellomo, Antonella Di Iulio, Achille Lococo, Annamaria Porreca, Piera Attilia Di Felice, and Raoul Saggini

Abstract

Lung cancer is one of the leading causes of cancer death worldwide. Surgical removal remains the best option for most tumors of this type. Reduction of cigarette consumption in patients with lung cancer candidates for the surgery could limit the impact of tobacco on postsurgical outcomes. Breathing exercises appear to help combat cigarette cravings. Yoga exercise benefits have been studied in lung cancer survivors, rather than in the preoperative setting. In this study, we have recruited 32 active smokers affected by lung cancer and being candidates for pulmonary surgery. The patients were randomly assigned to two groups: one treated by standard breathing and the other treated by yoga breathing (YB). The groups were evaluated at times T0 (baseline) and T1 (after 7 days of treatment) to compare the effects of the two breathing treatments on pulmonary performance in a presurgery setting. Pulmonary and cardiocirculatory functions have been tested using a self-calibrating computerized spirometer and a portable pulse oximetry device. The findings demonstrate appreciable short-term improvement in lung function assessed by spirometry. We conclude that yoga breathing can be a beneficial preoperative support for thoracic surgery.

Keywords

Active smokers · Breathing exercises · Lung cancer · Pranayama · Preoperative rehabilitation · Pulmonary function · Spirometry · Thoracic surgery · Yoga

1 Introduction

Lung cancer is one of the leading causes of cancer death worldwide. Surgical removal remains the best option for patients with Stage I and II of non-small cell lung cancer (NSCLC) and for selected patients with locally advanced disease (Stage IIIA) (Sisk and Fonteyn 2016). Approximately 20–30% of patients affected by lung cancer, candidates for pulmonary resection, are past or current smokers at diagnosis. The majority of patients who receive a lung cancer diagnosis

G. Barassi (✉), A. Porreca, P. A. Di Felice, and R. Saggini
Department of Medical Oral and Biotechnological Science, "Gabriele d'Annunzio" University, Chieti-Pescara, Italy
e-mail: coordftgb@unich.it

R. G. Bellomo
Department of Biomolecular Sciences, 'Carlo Bo' University, Urbino, Italy

A. Di Iulio and A. Lococo
Division of Thoracic Surgery, Santo Spirito Hospital of Pescara, Pescara, Italy

report that they are smokers and want to stop smoking (Morgan et al. 2011). The tobacco dependence is probably the most difficult dependency to break and often requires repeat interventions and attempts to quit (Peddle et al. 2009).

A reduction of cigarette consumption in hospitalized patients with lung cancer awaiting surgery is essential as inhaling of tobacco smoke has a negative effect on postsurgical outcome, due to carbon monoxide and nicotine cigarette content, which increases heart rate, blood pressure, and body's demand for oxygen (Dressler et al. 1996). Nicotine also causes vasoconstriction reducing the blood flow to certain parts of the body (Fiore 2008). Moreover, smoking causes the intrapulmonary small airways to narrow making them more prone to collapse and leading to increased propensity for infection, cough, pulmonary complications, and prolonged mechanical ventilation (Rejali et al. 2005).

Breathing exercises have been proposed as a way of combating cigarette cravings, potentially presenting a low-cost, easily scalable smoking cessation aid. In fact, a focus of attention on breathing and on physical sensations associated with inhalation and exhalation, which partially mimic cigarette smoking, may reduce the urge to smoke. Further, breathing may help smokers to relax and to counteract common withdrawal symptoms (Ngaage et al. 2002). Recent studies have mainly investigated the benefits of yoga exercises in lung cancer survivors (Sisk and Fonteyn 2016), but there is a lack of evidence about the benefit of yoga practice in the lung cancer preoperative setting.

The practice of yoga, meaning "union" the union of mind, body, spirit, and "the oneness of all things," originated about 5000 years ago (Katiyar and Bihari 2006). This practice, as other meditation techniques (Pokorski and Suchorzynska 2018), combines physical posture, breathing exercises, meditation, and philosophy. Yoga is an age-old traditional Indian psycho-philosophical-cultural method of leading one's life, which alleviates stress, induces relaxation, and provides multiple health benefits to the practicing person. It is a method of controlling the mind through the union of an individual's dormant energy with the universal energy. The commonly practiced yoga methods are "pranayama" (deep breathing), "asanas" (physical postures), and "dhyana" (meditation) admixed in varying proportions with differing philosophical background (Katiyar and Bihari 2006).

Recent studies have shown a beneficial role of yoga exercises in the management of health problems, for instance, asthma (Cooper et al. 2003), depressive disorders (Jayatunge and Pokorski 2018), and chronic obstructive pulmonary disease (COPD) (Kaminsky et al. 2017). Yoga breathing produces positive effects in patients with moderate-to-severe COPD, assessed from the improvement of the lung function variables and increments in exercise tolerance and psychological function (Sharma et al. 2005).

Early studies indicate that yoga and above all pranayama are feasible and well tolerated in patients with respiratory pathologies, and these meditation techniques may decrease dyspnea, increase exercise capacity, and improve oxygenation and quality-of-life scores. Focusing on body, mind, and spirit, yoga could be particularly helpful for individuals coping with cancer. In fact, yoga treatment can produce improvements also in the psychological functioning, with less anxiety and depression and increased feeling of hope, control, and self-esteem (Katiyar and Bihari 2006). This kind of psychological enhancement is of values in the face of depression and anxiety oftentimes accompanying cancer (Brown and Kroenke 2009). Further, yoga breathing focuses on the control of respiratory timing. The use of respiratory timing, expressed in seconds, is crucial because it allows the patient to do the best in each breath and allows the physiotherapist to manage the inhalation, air retention, and exhalation in the optimum way. Hatha yoga has been proposed as an adjunctive preoperative or perioperative therapy to improve patients' exercise capacity and to decrease the risk of postoperative pulmonary complications (Shahab et al. 2013). Although preoperative pulmonary rehabilitation has been widely suggested as an intervention to reduce surgical morbidity, an established treatment protocol does not exist (Shannon 2010).

Standard pulmonary rehabilitation uses deep breathing to improve exercise capacity and relaxation. Deep breathing slows respiratory rate, which leads to longer exhalation, better lung emptying, and a reduction in dynamic hyperinflation. Compared to yoga breathing, deep breathing is not involved with respiratory timing changes but rather focuses on breathing mechanics. In contradistinction, the breathing technique of pranayama controls every phase of the breath, which is important for reducing anxiety and withdrawal symptoms before thoracic surgery (Benzo et al. 2012; Holloway and Ram 2004).

Although medical benefits of yoga breathing are recognized, there is an apparent lack of studies on the use of yoga treatment in the lung cancer preoperative setting (Fouladbakhsh et al. 2014). Therefore, this study seeks to compare standard pulmonary rehabilitation with yoga breathing rehabilitation as treatment to improve pulmonary performance in smokers with lung cancer in a presurgery setting.

2 Methods

2.1 Study Design

The study was conducted by the Faculty of Physical and Rehabilitation Medicine of the "G. d'Annunzio" University of Chieti-Pescara in collaboration with the Division of Thoracic Surgery of the "Santo Spirito" Hospital in Pescara (Italy). All the patients gave informed written consent to the experimental procedure, which was in accordance with the latest revision of the Helsinki Declaration for Human Research and with the procedures concerning the privacy protection of subjects participating in biomedical research, as defined by ISO 9001 standards for research and experimentation. The study was approved by the institutional ethics committees.

Thirty-two current smokers (25 males and 7 females, Caucasians) diagnosed with primary NSCLC lung cancer in Stages I and II, candidates for surgical tumor removal that was judged the best curative option, were selected for the study. They were randomly assigned to two groups of 16 participants each: treated by standard breathing (SB) and treated by yoga breathing (YB). The groups were evaluated at two presurgery time points: T0 (baseline) and T1 (after 7-day-long rehabilitation treatment). The inclusion criterion was forced expiratory volume in 1 s (FEV_1) >60% predicted, which gives a low risk of perioperative mortality and respiratory morbidity (Peddle et al. 2009; Licker et al. 2006). The exclusion criteria for both groups were as follows: no eligibility for thorax surgery or metastases, acute diseases or health problems, and comorbidities incompatible with the study protocol such as neurological or cardiovascular diseases.

2.2 Study Procedures and Outcome Measures

All the patients were subjected to medical examination to delineate the characteristics of lung function, the presence and severity of lung cancer, and the eligibility to thoracic video-assisted surgery. This type of surgical approach was chosen since it is associated with a better preservation of clinical condition and a lower requirement for analgesics in the early postoperative period, compared with the standard surgery approach (Karasakia et al. 2009).

Pulmonary function was tested using a Spirolab III (MIR Medical International Research, Roma, Italy), a self-calibrating computerized spirometer that fulfills the criteria for standardized lung function tests. The subject was instructed to take maximum inspiration and blow into the mouthpiece as rapidly, forcefully, and completely as possible. A tight seal was maintained between lips and mouthpiece. Patients performed three tests in each evaluation, and the best of the three was taken into account for further analysis. The following outcome measures were evaluated; all expressed as % of predicted values:

- Forced vital capacity (FVC) – the amount of air that can be forcibly exhaled from the lungs after taking the deepest breath possible. FVC $\geq$ 80% predicted, calculated from age, height, weight, gender, and ethnic group, was considered the norm.

- Forced expiratory volume in 1 s (FEV$_1$) – the amount of air exhaled in 1 s, which was taken as a measure of airflow limitation. FEV$_1$ $\geq$ 80% predicted was considered the norm.
- FEV$_1$/%FVC ratio – the Tiffeneau-Pinelli index. The ratio $\geq$ 70% predicted was considered the norm (Ardestani and Abbaszadeh 2014).
- Peak expiratory flow (PEF) – the maximal flow achieved during the maximum forced expiration started after a full inspiration.
- Peak inspiratory flow (PIF) – the maximum instantaneous flow achieved during a forced inspiration started after a full expiration (Miller 2004).

In addition, peripheral oxygen saturation (SpO$_2$) of capillary blood hemoglobin and heart rate (HR) were assessed in each patient using a portable pulse oximeter (Nellcor™ Portable SpO$_2$ Patient Monitoring System PM1 0 N, Medtronic Inc., Dublin, Ireland).

2.3   Intervention Protocol

The YB patients practiced breathing under the supervision of a physiotherapist (Shahab et al. 2013). Treatment consisted of inhalation by first expanding the abdomen and then the chest using one slow and uninterrupted movement, until the maximum possible amount of air was drawn into the lungs, followed by a passively exhalation, accompanied by a feeling of relief and relaxation. The respiratory movement should be a wave-like, continuous abdomen-to-chest motion (Bellomo et al. 2012). Respiration should be accompanied by a sense of calmness. In addition, patients were asked to focus on extending expiration to counteract air trapping in the lungs and to slow expiratory flow. Breathing cycle was timed to 12 s and the timeline of breathing pattern was as follows: 4 s of inspiration, 4 s of air retention, and 8 s of expiration. The procedure was separated into

three sets of 10 yoga breaths each, interspersed with 30–60 s pauses between each set, and it was repeated 30 times a day.

The SB patients performed deep and slow breathing. This practice consisted of 30 repetitions of deep breathing, with a deep inspiration exploiting the abdomen and slow expiration through half-closed lips. Akin to the YB group, breathing exercise included 3 sets of 10 deep breaths, interspersed with 30–60 s pauses, and it was repeated 30 times a day.

2.4   Statistical Evaluation

Distribution of data at baseline was assessed with the Shapiro-Wilks normality test. Since the majority of variables did not pass the normality test, nonparametric analysis was employed. Medians of each variable were calculated for both YB and SB groups of patients in preparation for lung cancer surgery and were compared with the Mann-Whitney U test. Differences in the effectiveness of YB and SB were compared at two time points, after (T1) versus before (T0) therapeutic intervention with the Wilcoxon signed-rank test. Correlations between variables were tested with the Pearson method. The level for significance was set at $p < 0.05$. All tests were performed using the R statistical software.

3   Results and Discussion

Table 1 displays the descriptive evaluation of pulmonary function in the YB and SB groups. The patients of both groups were age-matched and also were matched from the clinical standpoint of the existing lung cancer. The inspection of the table shows that the mean values of breathing variables improved across the board in the YB group of patients, as opposed to the SB group where no real improvement could be discerned. The significance of the improvement in the YG, but not SB, group is clearly confirmed in the statistical evaluation of the increases in the values of individual pulmonary function variables from T0 (baseline) to T1 (after 7-day-long treatment) in

Table 1 Evaluation of pulmonary function in the yoga and standard breathing groups of lung cancer patients in preparation for surgery

Variables	Yoga breathing (YB) ($n = 16$)		Standard breathing (SB) ($n = 16$)	
	Mean ± SD	Min – max	Mean ± SD	Min – max
Age (year)	71.3 ± 7.0	59–82	71.8 ± 7.0	58–81
FVC T0 (%)	83.1 ± 15.6	58–113	92.2 ± 16.8	63–121
FVC T1 (%)	95.0 ± 15.6	66–124	94.0 ± 17.2	66–124
FEV_1 T0 (%)	75.7 ± 10.6	62–90	93.1 ± 20.6	64–124
FEV_1 T1 (%)	93.4 ± 16.2	75–125	90.9 ± 20.5	57–121
TIFF T0 (%)	56.2 ± 7.2	42.9–69.4	61.8 ± 8.1	41.8–74
TIFF T1 (%)	60.1 ± 6.7	50.2–72.0	59.2 ± 7.9	35.8–69.3
PIF T0 (%)	35.3 ± 16.1	16–70	41.1 ± 22.9	19–84
PIF T1 (%)	45.8 ± 17.5	23–80	43.4 ± 27.3	10–112
PEF T0 (%)	56.1 ± 20.0	23–86	70.3 ± 27.8	38–125
PEF T1 (%)	69.9 ± 19.2	30–96	65.3 ± 31.2	25–121
HR T0 (bpm)	79 ± 7	73–95	77 ± 5	70–85
HR T1 (bpm)	65 ± 4	56–73	68 ± 2	65–72
SpO_2 T0 (%)	94.8 ± 1.0	94–96	95.9 ± 0.6	95–97
SpO_2 T1 (%)	99.6 ± 1.0	97–99	96.8 ± 0.6	96–98

FVC forced vital capacity, *FEV_1* forced expiratory volumes 1 s, *TIFF* Tiffeneau-Pinelli index, *PIF* peak inspiratory flow, *PEF* peak expiratory flow, *HR* heart rate, *SpO_2* oxygen saturation; T0 (baseline); and T1 (after 7-day-long rehabilitation during presurgery period)

Table 2 Comparison of the significance of changes in breathing variables from T0 (baseline) to T1 (after 7-day long treatment) within the yoga breathing (YB) group and standard breathing (SB) group

Variables	YB T0 vs. T1 p-value	SB T0 vs. T1 p-value
FVC	<0.0001	0.801
FEV_1	<0.0001	0.146
TIFF	0.0130	0.003
PIF	<0.0001	0.820
PEF	<0.0001	0.220
HR	<0.0001	0.0001
SpO_2	<0.0001	0.0020

Abbreviations are same as in Table 1; Wilcoxon signed-rank test for paired data

either group of cancer patients displayed in Table 2.

A significant improvement in FVC and FEV_1 from the baseline T0 to the after treatment T1 in the YB, but not SB, group of patients is also seen when the median values are considered. It is noticeable that the medians of FVC and FEV_1 were lower in the YB than SB group at T0

(Fig. 1), which explains the lack of a significant difference in either variable between the YB and SB group at T1 (Table 3), as the variables increased toward the SB level over time of training while the SB did not. Concerning the PIF and PEF variables, both significantly increased in the YB group, while there was just an increasing trend in the SG group (Figs. 1 and 2, Table 2). Nonetheless, the differences remained insignificant between the two groups (Table 3), which suggest that the improvement was temporary only. Mutual relationships between respiratory variables became strengthened after yoga breathing training, compared with the SB group, as judged from Pearson's correlation coefficients (Table 4), which may be taken as a greater stabilization of the breathing pattern.

Of note, cardiac function, expressed by stabilization of HR and increase in SpO_2, improved significantly from T0 to T1 in both groups (Table 2); the improvements were distinctly greater in the YG group of patients (see details in Tables 1, 2, and 3 and Figs. 1 and 2). The increase in SpO_2 became related with that in FEV_1 after yoga breathing training, the

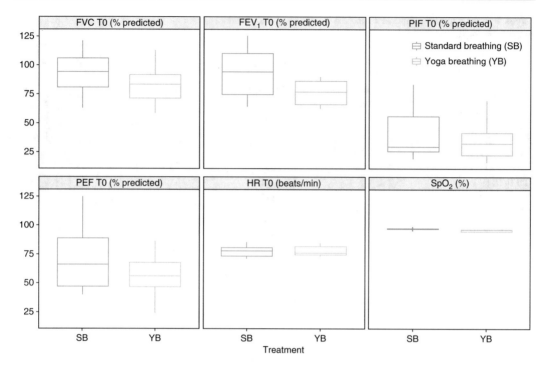

Fig. 1 Box plots of the variables studies at T0 (baseline before treatment). *FVC* forced vital capacity, *FEV₁* forced expiratory volumes 1 s; PIF, peak inspiratory flow, *PEF* peak expiratory flow, *HR* heart rate; SpO₂, oxygen saturation. Median and interquartile range (IQR) is shown; $IQR = Q3 - Q1$. Bars indicate the outliers outside $Q1–1.5IQR$, $Q3 + 1.5IQR$

Table 3 Comparison of the significance of differences in individual breathing variables at T1 time (after 7-day-long treatment) between the yoga breathing (YB) and standard breathing (SB) groups

Variables	p-value
FVC T1	0.909
FEV₁ T1	0.748
TIFF T1	0.780
PIF T1	0.497
PEF T1	0.571
HR T1	0.021
SpO₂ T1	<0.0001

Abbreviations are same as in Table 1; Mann-Whitney U test for unpaired data

association not present in the SB group (Table 4). The improvement of cardiac function reported in the present study constitutes an added value to the results of previous studies that have demonstrated a better lung function after the yoga breathing training (Yadav et al. 2009).

The essential findings of this study were that pranayama treatment caused a significant increase in FEV₁ and FVC from T0 to T1 measurement points, i.e., baseline vs. after 7-day-long rehabilitation treatment, in the yoga breathing group of primary lung cancer patients scheduled for thoracic. This increase was absent in the standard breathing group of patients. Respiratory improvement was, in all likelihood, due to slow deep breathing with breath holding and extended expiratory time with the involvement of abdominal expiratory muscles, the features that help relieve gas trapping (Dechman and Wilson 2004). Pranayama type of breathing also reduces deadspace ventilation and work of breathing and refreshes the air throughout the lungs, as opposed to superficial breathing that refreshes the air only at the lung base.

Yadav and Das (2001) have reported an improvement in respiratory functional variables in response to yoga breathing and have attributed the effect to increments in respiratory muscle strength, clearing of respiratory secretions, and using the diaphragmatic and abdominal muscles

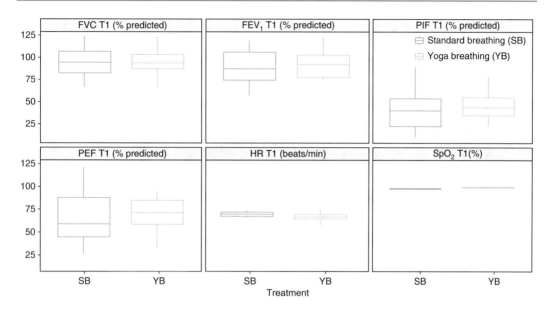

Fig. 2 Box plots of the variables studies at T1 (after 7-day-long treatment). *FVC* forced vital capacity, *FEV₁* forced expiratory volumes 1 s, *PIF* peak inspiratory flow, *PEF* peak expiratory flow, *HR* heart rate, *SpO₂* oxygen saturation. Median and interquartile range (IQR) is shown; $IQR = Q3 - Q1$. Bars indicate the outliers outside $Q1{-}1.5IQR$, $Q3 + 1.5IQR$

ancillary respiratory muscles for filling the respiratory structures more efficiently and completely. Moreover, increased strength of respiratory muscles due to yoga breathing enhances ventilation and perfusion of poorly ventilated parts of lungs, thereby improving oxygen delivery to tissues. These effects correct the imbalance of the ventilation-perfusion ratio and consequently may lead to a gain in the PIF value. Likewise, the present findings are in accord with the previous studies that show the PEF improves over the time of 2 weeks of pranayama practice (Yadav et al. 2009).

Pranayamic breathing, characterized by a regular and slow frequency respiration, with long periods of breath retention, is known to affect human physiology. One of the long-term effects of this way of breathing is improved autonomic function consisting of an increase in parasympathetic activity and a blunted sympathetic dominance (Singh et al. 2004). The short-term effects of yoga breathing, on the other hand, include a decrease in oxygen consumption and heart rate and blood pressure. Moreover, it has been suggested that the cardiorespiratory system is normalized through rhythmic breathing exercises, such as slow yoga breathing (Gopal et al. 1973). Thus, this pattern of breathing functionally resets the autonomic nervous system through stretch-induced inhibitory signals and hyperpolarizing currents that propagate through both neural and non-neural tissues, all of which helps synchronize neural elements of the heart, lungs, limbic system, and cortex (Peddle et al. 2009).

A strength of this study lies in the demonstration of a relationship between pranayama practice and improved lung function in active smokers with lung cancer scheduled for thoracic surgery. The study also shows that yoga breathing is superior to standard respiratory training and it is applicable in the primary preoperative setting and pulmonary rehabilitation programs. Yoga breathing training would be particularly useful in the home of patients not included in a formal program of pulmonary rehabilitation. Further, yoga breathing could be a useful aid to smoking cessation. In terms of limitations, the present findings cannot be extended to the postoperative status of patients. Also, further trials in a larger sample of patients should be performed to increase the

Table 4 Correlation coefficients between the pairs of variables studied in yoga breathing group (**a**) and standard breathing group (**b**) of lung cancer patients in preparation for surgery

(a)	Age	FVC T0	FVC T1	FEV$_1$ T0	FEV$_1$ T1	TIFF T0	TIFF T1	PIF T0	PIF T1	PEF T0	PEF T1	HR T0	HR T1	SpO$_2$ T0	SpO$_2$ T1
Age (yr)	1														
FVC T0	−0.027	1													
FVC T1	−0.207	0.894***	1												
FEV$_1$ T0	0.115	0.730**	0.697**	1											
FEV$_1$ T1	0.031	0.813***	0.870***	0.790***	1										
TIFF T0	0.088	−0.553*	−0.409	0.066	−0.159	1									
TIFF T1	0.303	0.011	−0.067	0.277	0.388	0.539*	1								
PIF T0	−0.164	−0.119	0.149	0.030	0.045	0.114	−0.212	1							
PIF T1	−0.240	−0.098	0.128	−0.007	−0.009	0.090	−0.211	0.912***	1						
PEF T0	−0.127	0.748***	0.776***	0.496	0.626**	−0.521*	−0.156	0.408	0.372	1					
PEF T1	−0.314	0.706**	0.804***	0.498*	0.574*	−0.442	−0.319	0.408	0.475	0.804***	1				
HR T0	−0.298	−0.267	−0.091	−0.278	−0.355	0.096	−0.479	0.182	0.327	−0.169	0.192	1			
HR T1	0.191	0.056	0.019	0.082	−0.086	−0.030	−0.200	0.186	0.182	0.222	0.315	0.293	1		
SpO$_2$ T0	0.365	0.108	0.056	0.422	0.299	0.242	0.306	−0.206	0.182	−0.129	−0.048	−0.296	−0.148	1	
SpO$_2$ T1	0.169	−0.290	−0.339	−0.292	−0.548*	−0.060	−0.514*	−0.087	0.016	−0.302	−0.082	0.485	0.487	−0.427	1
(b)	Age	FVC T0	FVC T1	FEV$_1$ T0	FEV$_1$ T1	TIFF T0	TIFF T1	PIF T0	PIF T1	PEF T0	PEF T1	HR T0	HR T1	SpO$_2$ T0	SpO$_2$ T1
Age (yr)	1														

	FVC T0	FVC T1	FEV$_1$ T0	FEV$_1$ T1	TIFF T0	TIFF T1	PIF T0	PIF T1	PEF T0	PEF T1	HR T0	HR T1	SpO$_2$ T0	SpO$_2$ T1
FVC T0	−0.443													
FVC T1	−0.677**	1												
FEV$_1$ T0	−0.258	0.865***	1											
FEV$_1$ T1	−0.349	0.805***	0.723**	1										
TIFF T0	0.212	0.699**	0.745***	0.956***	1									
TIFF T1	0.302	−0.129	−0.050	0.467	0.546*	1								
PIF T0	−0.150	−0.047	−0.107	0.510*	0.557*	0.931***	1							
PIF T1	−0.740**	0.370	0.245	0.480	0.443	0.290	0.420	1						
PEF T0	−0.517*	0.367	0.584*	0.127	0.190	−0.316	−0.400	0.409	1					
PEF T1	−0.346	0.639**	0.629**	0.760***	0.737**	0.310	0.308	0.672**	0.573*	1				
HR T0	−0.026	0.476	0.443	0.792***	0.792***	0.586*	0.603*	0.618*	0.289	0.913***	1			
HR T1	0.115	−0.321	−0.127	−0.393	−0.378	−0.402	−0.418	−0.363	−0.190	−0.473	−0.535*	1		
SpO$_2$ T0	0.023	0.086	−0.185	−0.161	−0.105	0.155	0.056	−0.339	−0.053	−0.115	0.046	−0.022	1	
SpO$_2$ T1	−0.050	0.287	−0.031	0.246	0.204	0.334	0.376	0.114	−0.242	0.130	0.344	0.024	0.280	1

Abbreviations are same as in Table 1, *0.05; **0.01; ***0.001.

statistical power of functional data analysis over time and reduce the potential type II error (Maturo and Di Battista 2018; Di Battista et al. 2017). Future direction of research on the effects on lung function of meditative breathing methods should combine manual neuromuscular therapy with yoga respiration as autonomic effects of manual therapy induce reflex rebalancing of the heart and respiratory rates, which modulates the whole body system and thus could help increase the effectiveness of yoga therapy (Barassi et al. 2017; Delli Pizzi et al. 2017).

4   Conclusions

This study contributes to the knowledge about the benefits of pranayama meditative practice on lung function in active smokers affected by lung cancer scheduled for surgery. Although standard breathing exercises are effective to an extent, yoga breathing is an alternative option that may provide the optimum short-term improvement in the lung function. Thus, yoga breathing may become a valid support for the preoperative thoracic surgery preparation. Further, yoga breathing holds a potential to help smokers quit and to improve their quality of life.

Competing Interests The authors declare no competing interests in relation to this article.

References

Ardestani ME, Abbaszadeh M (2014) The association between forced expiratory volume in one second (FEV1) and pulse oximetric measurements of arterial oxygen saturation (SpO$_2$) in the patients with COPD: a preliminary study. J Res Med Sci 19(3):257–611

Barassi G, Bellomo RG, Porreca A, Di Felice PA, Prosperi L, Saggini R (2017) Somato-visceral effects in the treatment of dysmenorrhea: neuromuscular manual therapy and standard pharmacological treatment. J Altern Complement Med; doi: https://doi.org/10.1089/acm.2017.0182

Bellomo RG, Barassi G, Iodice P, Di Pancrazio L, Megna M, Saggini R (2012) Visual sensory disability: rehabilitative treatment in an aquatic environment. Int J Immunopathol Pharmacol 25(1):17–22

Benzo R, Wigle D, Novotny P, Wetzstein M, Nichols F, Shen RK, Cassivi S, Deschamps C (2012) Preoperative

pulmonary rehabilitation before lung cancer resection: results from two randomized studies. Lung Cancer 74 (3):441–445

Brown F, Kroenke K (2009) Cancer-related fatigue and its associations with depression and anxiety: a systematic review. Psychosomatics 50(5):440–447

Cooper S, Oborne J, Newton S (2003) Effect of two breathing exercises (Buteyko and Pranayama) in asthma: a randomized controlled trial. Thorax 58:674–679

Dechman G, Wilson C (2004) Evidence underlying breathing retraining in people with chronic obstructive pulmonary disease. Phys Ther 84:1189–1197

Delli Pizzi S, Bellomo RG, Carmignano SM, Ancona E, Franciotti R, Supplizi M, Barassi G, Onofrj M, Bonanni L, Saggini R (2017) Rehabilitation program based on sensorimotor recovery improves the static and dynamic balance and modifies the basal ganglia neurochemistry: a pilot 1H-MRS study on Parkinson's disease patients. Medicine 96(50):8732

Di Battista T, Fortuna F, Maturo F (2017) BioFTF: an R package for biodiversity assessment with the functional data analysis approach. Ecol Indic 73:726–732

Dressler CM, Bailey M, Roper CR (1996) Smoking cessation and lung cancer resection. Chest 110:1199–1202

Fiore MC (2008) Treating tobacco use and dependence: 2008 update U.S. Public Health Service Clinical Practice Guideline executive summary. Respir Care 53:1217–1222

Fouladbakhsh JM, Davis JE, Yarandi HN (2014) A pilot study of the feasibility and outcomes of yoga for lung cancer survivors. Oncol Nurs Forum 41(2):162–174

Gopal KS, Anantharaman V, Balachander S, Nishith SD (1973) The cardiorespiratory adjustments in 'Pranayama', with and without 'Bandhas', in 'Vajrasana'. Indian J Med Sci 27(9):686–692

Holloway E, Ram FS (2004) Breathing exercises for asthma. Cochrane Database Syst Rev 1:CD001277

Jayatunge RM, Pokorski M (2018) Post-traumatic stress disorder: a review of therapeutic role of meditation interventions. Adv Exp Med Biol doi: https://doi.org/10.1007/5584_2018_167

Kaminsky DA, Guntupalli KK, Lippmann J, Burns SM, Brock MA, Skelly J, DeSarno M, Pecott-Grimm H, Mohsin A, LaRock-McMahon C, Warren P, Whitney MC, Hanania NA (2017) Effect of yoga breathing (Pranayama) on exercise tolerance in patients with chronic obstructive pulmonary disease: a randomized, controlled trial. J Altern Complement Med 23 (9):696–704

Karasakia T, Nakajimab J, Murakawab T, Fukamib T, Yoshidab Y, Kusakabeb M, Hiroshi O, Takamotob S (2009) Video-assisted thoracic surgery lobectomy preserves more latissimus dorsi muscle than conventional surgery. Interact Cardiovasc Thorac Surg 8:316–320

Katiyar SK, Bihari S (2006) Role of Pranayama in rehabilitation of COPD patients – a randomized controlled study. Indian J Allergy Asthma Immunol 20(2):98–104

Licker MJ, Widikker I, Robert J, Frey JG, Spiliopoulos A, Ellenberger C, Schweizer A, Tschopp JM (2006)

Operative mortality and respiratory complications after lung resection for cancer: impact of chronic obstructive pulmonary disease and time. Trend Ann Thorac Surg 81(5):1830–1837

Maturo F, Di Battista T (2018) A functional approach to Hill's numbers for assessing changes in species variety of ecological communities over time. Ecol Indic 84:70–81

Miller MR (2004) Peak expiratory flow meter scale changes: implications for patients and health professionals. Airways J 2(2):80–82

Morgan G, Schnoll RA, Alfano CM, Evans SE, Goldstei A, Ostroff J, Park ER, Sarna L, Sanderson-Cox L (2011) National Cancer Institute conference on treating tobacco dependence at cancer centers. J Oncol Pract 7(3):178–182

Ngaage DL, Martins E, Orkell E (2002) The impact of the duration of mechanical ventilation on the respiratory outcome in smokers undergoing cardiac surgery. Cardiovasc Surg 10:345–350

Peddle CJ, Jones LW, Eves ND, Reiman T, Sellar CM (2009) Effects of presurgical exercise training on quality of life in patients undergoing lung resection for suspected malignancy: a pilot study. Cancer Nurs 32 (2009):158–165

Pokorski M, Suchorzynska A (2018) Psychobehavioral effects of meditation. Adv Exp Med Biol 1023:85–91

Rejali M, Rejali AR, Zhang L (2005) Effects of nicotine on the cardiovascular system. Vasc Dis Prev 2:135–144

Shahab L, Sarkar BK, West R (2013) The acute effects of yoga breathing exercises on craving and withdrawal symptoms in abstaining smokers. Psychopharmacology 225:875–882

Shannon VR (2010) Role of pulmonary rehabilitation in the management of patients with lung cancer. Curr Opin Pulm Med 16:334–339

Sharma VK, Das S, Mondal S, Goswampi LJ, Gandhi A (2005) Effect of Sahaj Yoga on depressive disorders. Indian J Physiol Pharmacol 49:462–468

Singh S, Malhotra V, Singh KP, Madhu SV, Tandon OP (2004) Role of yoga in modifying certain cardiovascular functions in type 2 diabetic patients. J Assoc Physicians India 52:203–206

Sisk A, Fonteyn M (2016) Evidence-based Yoga interventions for patients with cancer. Clin J Oncol Nurs 20(2):181–186

Yadav RK, Das S (2001) Effect of yoga practice on pulmonary functions in young females. Indian J Physiol Pharmacol 45:493–496

Yadav A, Savita S, Singh KP (2009) Role of the Pranayama breathing exercises in rehabilitation of coronary artery disease patients – a pilot study. Indian J Tradit Knowl 8(3):455–458

Adv Exp Med Biol - Clinical and Experimental Biomedicine (2018) 1: 31–39
https://doi.org/10.1007/5584_2018_192
© Springer International Publishing AG, part of Springer Nature 2018
Published online: 31 March 2018

Effects of Breast and Prostate Cancer Metastases on Lumbar Spine Biomechanics: Rapid In Silico Evaluation

J. Lorkowski, O. Grzegorowska, M. S. Kozień, and I. Kotela

Abstract

Metastases to distant organs are a frequent occurrence in cancer diseases. The skeletal system, especially the spine, is one such organ. The objective of this study was to apply a numerical modeling, using a finite element method (FEM), for the evaluation of deformation and stress in lumbar spine in bone metastases to the spine. We investigated 20 patients (10 women and 10 men) aged 38–81 years. In women, osteolytic lesions in lumbar spine accompanied breast cancer, in men it was prostate cancer. Geometry of FEM models were built based on CT scans of metastatic lumbar spine. We made the models for osteolytic metastases, osteosclerotic metastases, and metastases after surgery. Images were compared. We found a considerable concentration of strain, especially located in the posterior part of the vertebral body. In osteolytic lesions, the strain was located below the vertebral body with metastases. In osteosclerotic lesions, the strain was located in the anterior and posterior parts in and below the vertebral body with metastases. Surgery abolished the pathological strain. We conclude that metastases to the lumbar spine introduce a pathological strain on the lumbar body. The immobilization of the vertebral body around fractures abolished the strain.

Keywords

Cancer metastases · Finite element method · Lumbar spine · Osteolytic lesions · Osteosclerotic lesions · Strain

1 Introduction

Malignancy is the main therapeutic challenge of modern medicine, alongside the cardiovascular system diseases. Although the prevalence of deaths resulting from cancer has declined in the decade of 2004–2013, cancer remains the most frequent cause of death. The postcranial skeletal system is the third, after the lungs and liver, most common site of metastases, with the spine being affected in 61.8% (Pawlak et al. 2002). Bone malignancies become a rare direct cause of death, but they definitely cause disability and poor quality of life.

J. Lorkowski (✉)
Department of Orthopedics and Traumatology, Central Clinical Hospital of Ministry of Interior, Warsaw, Poland

Health Rehabilitation Center, Cracow, Poland
e-mail: jacek.lorkowski@gmail.com

O. Grzegorowska
Health Rehabilitation Center, Cracow, Poland

M. S. Kozień
Department of Material Systems Dynamics, Faculty of Mechanical Engineering, Institute of Applied Mechanics, Cracow University of Technology, Cracow, Poland

I. Kotela
Department of Orthopedics and Traumatology, Central Clinical Hospital of Ministry of Interior, Warsaw, Poland

Spinal metastases mostly result from lung cancer (31%), breast cancer (24%), gastrointestinal cancer (9%), prostate cancer (8%), lymphoma (6%), and renal cancer (1%) (Phanphaisarn et al. 2016). Bone malignancy can often manifest as the very first symptom of cancer (Kimbung et al. 2015). For instance, 5–10% of breast cancer patients already suffer from bone metastases at the moment of diagnosis (Ciftdemir et al. 2017).

There are different diagnostic modalities for spine pathologies. In clinical practice, X-ray can be used as the first step, in which an "owl eye" sign, i.e., an obliterated structure of the vertebral arch, can be seen. Osteoblastic metastases are much more difficult to picture, and the conventional X-ray should be supplemented with other diagnostic methods (Guzik 2016). Currently, MRI and PET/CT are considered the gold standard, mostly because of their high sensitivity and specificity. Nonetheless, some spinal lesion may have an ambiguous or blurred appearance due to an overlapping pathology of surrounding tissues or other reasons and thus be hard to diagnose in a firm manner (Kintzelé and Weber 2017). Clinicians often evaluate the biomechanical capacity of the spine based on CT scans. Scintigraphy, having a high sensitivity but low specificity, is used as a complementary diagnostic tool method. Other methods, such as positron emission tomography and spinal angiography, can be harnessed to obtain a better diagnostic yield (Ecker et al. 2005).

Symptoms presented by patients with the spinal metastatic disease are mostly pain and neurological disorders. In the majority, pain appears on palpation or percussion of a pathological tissue or site (Pawlak et al. 2002). Neurological deficits are most likely to be seen among patients with metastases to the thoracic spine, which can put a direct pressure on the spinal cord. Moreover, breathing, urination, and defecation disorders, Horner's syndrome, Brown-Sequard syndrome, Beevor's sign, and general signs of weakness, appetite and weight loss, or herpes zoster can be observed (Kopaczewski et al. 2009; Jacobs and Perrin 2001).

Qualification for treatment is multifactorial. There are a great deal of scoring classifications and scales used to assess the spine stability, and prognostic and therapeutic factors in spinal cord injury, the exemplary of which are Denis (1983), Asdourian et al. (1990a), Tomita et al. (2001), Tokuhashi et al. (2005), or Frankel's grade classification (Ditunno et al. 1994; Mc Cormaek et al. 1994). Based on therapeutic recommendations, patients are qualified to the valid treatment scheme, including surgery. Manifested symptoms, reduced quality of life, score results, and life-threatening conditions often prompt a decision of surgical intervention, which raises a need for insightful evaluation of biomechanics, and strain and strength vectors affecting the spine. The objective of this study was to assess a novel numerical modeling using finite element method (FEM) for the evaluation of deformation and stress in lumbar spine metastases.

2 Methods

The study was approved by the institutional Review Board for Research and was conducted in accordance with the Declaration of Helsinki for Human Research. We evaluated the distribution of mechanical stress in the spinal column, caused by metastatic strain and injury, in 2D in silico models prepared for 20 patients (10 women and 10 men) of the mean age of $67.0 \pm 12.8(SD)$ years (range 38–81 years). The women group consisted of patients diagnosed with breast cancer with osteolytic lesions, and the men had prostate cancer with osteosclerotic metastases to the lumbar spine. Patients of both groups had metastatic lesions in the lumbar spine and had pain corresponding with the topography of lesions. The spine lesions were evaluated in detail in anteroposterior and lateral X-rays and CT scans of lumbar spine. For in silico analysis, a segmental spine model was chosen, which encompassed the L2 vertebral body and two conterminous L1 and L3 vertebrae, in a patient with breast cancer and destruction of L2. The FEM was developed consisting of numerical algorithms for the simulation of physical vertebral phenomena, in which local mechanical strains resulting from metastatic injury were evaluated.

FEM is at present a key method of strength analysis of engineering structures. The method helps resolve the problems emerging in solid mechanics theory. In FEM modeling, the structure considered is represented by a sum of subareas (subvolumes) called the "finite element". These elements are connected at specified points called "nodes". At these points, the unknown displacements are determined, on basis of which the values of strain and stress are obtained. The actual values of these parameters inside the structural elements are approximated by a suitable function. In recent years, FEM is increasingly used to model the biosolid structures. The bones are natural biosolid structures that can be described by a solid structure model used for in silico analysis.

To evaluate the spine segments with metastatic lesions, Ansys computer software was used (Ansys Inc., Canonsburg, PA). Based on radiological images, a geometry model of tissues forming a spine segment with lesions and the two adjacent segments was created. There are four models included in the analysis. Firstly, an image compatible with clinical and radiological picture of a single osteolytic spine lesion caused by breast cancer was analyzed. In the second model, material properties were changed those typical for osteosclerotic metastases of prostate cancer. The lesion size was analogous to the osteolytic lesion of the first model. The other two models encompassed the conterminous vertebrae below and above the one with the metastatic lesion and presented a reposited fracture, as it were in the case of surgical treatment, or a fracture secured with Jewett's corset. Surgically reposited models were made based on models made for osteolytic and osteosclerotic fractures. The images were processed using an in silico modeling approach that combines dynamic simulation and FEM to describe the spine cancer-specific musculoskeletal dynamics. For the FEM model, CT images in midsagittal plane, obtained from the breast cancer patient with an osteolytic spine fracture, were used. CT2FEM 1.0 software applied to automatically transform 256 grayscale bitmaps into FEM model (Lorkowski et al. 2014). The application assigns 256 material features to

the FEM model based on the analysis of grayscale bitmaps. After importing the model into the Ansys computing system, the finite elements were grouped into 16 components (sets), presented as a contour line map, based on their material features. The areas representing spongy and compact bone, fibrous cartilage of the annulus of the intervertebral disc, the nucleus pulposus of the intervertebral disc, connective tissue of ligaments, and the neoplastic tissue were defined using the Boolean operations on the defined components of the FEM model. For numerical investigations, a simplified, homogeneous and isotropic description of tissue properties was assumed. A division of three vertebrae with a neoplastic change and a nucleus pulposus is shown in Fig. 1. The colors correspond to materials with different modeling properties: the bone, annulus fibrosis, nucleus pulposus, ligaments, and a metastatic tumor.

Stress and strain exerted on the spinal structures can be described by Young's modulus and Poisson's ratio of solid material's elasticity. The former is the ratio between stress and strain of a material, i.e., force applied to a material and the resulting material's elastic deformation, and the latter is the ratio between transverse and axial strain. Young's modulus (E) amounted to 5 MPa for compact bone, 3 MPa for spongy bone, 500 MPa for fibrous cartilage, 300 MPa for nucleus pulposus, 400 MPa for connective tissue of ligaments, 400 MPa for osteolytic metastases, and 4500 MPa for osteosclerotic metastases. Poisson's ratio (ν) was 0.32 for all the above outlined. A construction, consisting of the anatomic structures and the neoplastic lesion, modeled as an assembly of finite elements representing different regions, could be switched on/off as empty/full subdomains during the analysis (Lorkowski et al. 2014). Anatomically unitary entities in the representative cross section of the spine forming a 2D model were treated as being homogeneous (Wirtz et al. 2000). The FEM model consisted of one layer of eight-node solid 45 3D elements. It encompassed three vertebral bodies connected with spinous processes; the connection was modeled as a coupling of displacement of suitable nodes (Fig. 2). The first

Fig. 1 Division of three lumbar vertebrae with neoplastic change is shown in colors corresponding to different modeling properties of tissues: spongy bone (pink), compact bone (light blue), annulus fibrosus (green), nucleus pulposus (orange), ligaments (red), metastatic tumor (navy blue)

Fig. 2 Division of three lumbar vertebral bodies into elements with spinous processes and ligaments: couplings (green lines between the posterior sides of vertebral bodies and the bases of spinous processes), external loadings (red lines on the left-hand side), and boundary conditions (turquoise shade around the whole structure)

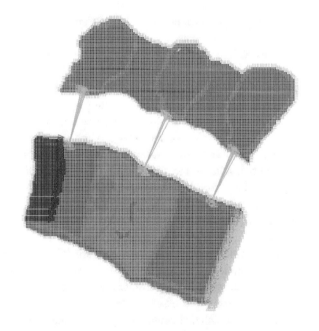

vertebral body was loaded with a force of $F_x = 343$ N, distributed on the external vertebral surface and acting in the horizontal plane (Bergmann et al. 2001). The mesh of elements together with visualization of external loads, boundary conditions, and couplings between the posterior parts of lumbar vertebral bodies and the spinous and articular processes is shown in Fig. 2. The results of FEM analysis are visualized as a map of equivalent stresses in accordance with von Mises's hypothesis for a combination of the elements forming three successive lumbar vertebrae. We employed this method in an attempt to model the abolition of lumbar spine movability above and below the fractured vertebral body.

3   Results

The basic parameter that is subject to comparative analysis is the distribution of the equivalent von Mises stress. The boundary conditions and external loadings are the same for all cases. Stress distribution in the non-cancer case is, in general, uniform for three vertebral bodies, except for some small areas with the apparent influence of boundary conditions and local couplings. A comparison of the distribution of the equivalent von Mises stress in the case of cancerous changes in the vertebral bodies of the lumbar spine is shown in Fig. 3a and b for osteolytic (breast cancer) and osteosclerotic (prostate cancer) lesions, respectively. Stress distribution was qualitatively the same in both, but its maximum level was greater in the osteolytic lesions. In the case of both lesion types, the maximum stress reached a reasonably realistic value, which was due likely to the presence of the single solid deformable body (bone) without boundary conditions.

The distribution of the equivalent von Mises stress for the whole analyzed cross section in both osteolytic and osteosclerotic cancer lesions is compared in Fig. 3c and d. The stress was distributed uniformly (dark blue), with relatively low values of up to about 140 MPa, and was qualitatively alike. A higher value of stress (light blue) and a concentration of high stress (yellow, orange, and red) resulted from couplings and boundary conditions existing in the model. Extremely high stress values, reaching 1261 MPa, are connected with the FEM approximate analysis and are not realistic. Such high levels could be reduced by a higher mesh discretization and employing nonrigid boundary conditions.

Stress concentration in the posterior part of the lumbar vertebrae, particularly located below the metastatic fractures, was noted in osteolytic lesions. Osteoblastic changes caused stress concentration not only in these posterior parts but also on the anterior edge of the fractured vertebra.

The FEM model presented corresponds to clinical stabilization of stress, which was achieved with the use of coupling. Stress concentration was suppressed in both osteolytic and osteoblastic fractures.

4   Discussion

Localization of spinal metastases results from several factors. One of them is the anatomic makeup, which may be exemplified by breast cancer. The mammary gland venous vessels are connected with the posterior and anterior external vertebral venous plexuses. This is the closest way through which neoplasmatic breast cells can migrate (Coleman 2006; Batson 1942). Paget's theory of "seed and soil" states that particular cells with specific properties (seed) can settle down and grow in particular conditions (soil) (Paget 1889). Contemporary science shows that syndecan-1, a transmembrane heparan sulfate proteoglycan that is essential for cell binding and cell signaling, may initiate the formation of metastases (Sanderson and Yang 2008). Exosomes and biological fluids play the role of transporters (David Roodman and Silbermann 2015). The products released from neoplasmatic cells and adhesive molecules additionally create a favorable microenvironment of the bone marrow to start the growth (Shiozawa et al. 2011).

There is a homeostatic balance between osteogenesis and resorption in the healthy bone. This balance is disrupted in metastatic lesions, leading to osteolytic or osteoblastic changes. Osteolytic lesions are commoner and fairly typical for breast cancer or myeloma multiplex, while osteoblastic ones are mostly observed in prostate, thyroid, bladder, and gastrointestinal cancers. There are a lot of mediators taking part in the metastatic process. Expression of parathyroid hormone-related protein (PTHrP) and of the molecular triad consisting of receptor activator of NF-kappaB ligand (RANKL), its receptor RANK, and osteoprotegerin (OPG), the endogenous soluble

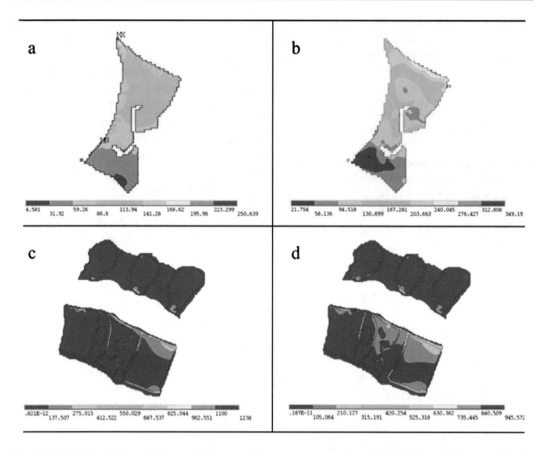

Fig. 3 Distribution of the equivalent von Mises stress in the region of cancerous lesions: (**a**) osteolytic lesion of breast cancer, (**b**) osteosclerotic lesion of prostate cancer, and for the whole three-vertebral cross section with cancerous changes (**c**) osteolytic and (**d**) osteosclerotic. The value of stress intensity increases with colors from left to right

RANKL inhibitor, tilts the homeostatic bone balance toward osteolysis (Dougall and Chaisson 2006), being a frequent cause of spine instability and deformation. On the other hand, endothelin, plasminogen activator, insulin-like growth factors, transforming growth factor beta (TGF-β), and integrins favor osteogenesis (Groblewska et al. 2008). Of note, PTHrP can be a product of prostate cancer cells and initiate the osteoblastic remodeling of bone tissue (Liao et al. 2008), which makes it of essential importance to consider this cancer in the differential diagnosis of osteolytic lesions. The influence of glucagon-like peptide-1 (GLP-1) on bone metabolism remains a contentious issue. Despite some evidence for the beneficial effect of GLP-1 on bone tissue regeneration, there also are arguments

to the contrary (Zhao et al. 2017). In general, bone remodeling is detrimental for biomechanics of the axial skeleton, because the newly formed bone elements cause immaturity of the reticular bone structure and increase vertebral vulnerability to damage (Hauge et al. 2001). As a result compression fractures may occur. Metastases to the axial skeleton lead to spine instability, escalation of pain, and neurological deficits caused by spinal cord compression. In the FEM model employed in the present study, differences in the spine biomechanics caused by osteolytic and osteoblastic lesions were shown in three representative lumbar spine segments. We found a concentration of stress in the affected lumbar vertebrae, which is typical for pain manifestation. In the case of breast and prostate cancer, bisphosphonates are recommended to decrease pain and prevent further pathological fractures (Wilson et al. 2012). Other recommended treatment modalities are conservative, consisting of steroids and radiotherapy or invasive surgery (Loblaw et al. 2012). Radiotherapy combined with surgery has a greater effectiveness than radiotherapy alone. The basic goal of treatment is to achieve spine decompression and stabilization. Surgical stabilization is recommended in radio-insensitive patients who remain in a good general condition and have a life expectancy of at least 3 months. Metastases that lead to neurological deficits are mostly located in the anterior part of the spinal (about 80% of cases), and always enfold the vertebral body, encompassing the pedicle of the vertebral arch in about half the cases. Only do one fifth of patients with lesions in the anterior part of the spine not suffer from spinal cord compression (Kato et al. 2012; Asdourian et al. 1990b).

Radiotherapy is an effective method in decreasing symptoms of neurological deficits and reducing pain severity. When symptoms last for less than 24 h and the tumor is radio-insensitive, a surgical decompression and stabilization with adjuvant radiotherapy should be considered (Bartels et al. 2008). Radiotherapy performed in palliative care increases the survival rate to 26% versus 14% in non-radiation patients with advanced breast cancer (Steinauer et al. 2014). Survival of patients from the time of a diagnosis of spine metastases differs and depends on the type of cancer. The average life expectancy is about 10 months (Coleman 2006). The presence of bone lesions is a sign of the advanced disease, which is why any increase in quality of life is of importance. Although the survival rate in the case of spinal canal metastases remains very poor, radiotherapy might maintain a degree of agility in symptomatic patients (Wilson et al. 2012). Radiotactic stereosurgery is expected to be used as adjuvant therapy after decompression in the near future. With this method, it should be possible to precisely define a proper surgical approach for optimum outcomes (Pokhrel et al. 2017; Di Martino et al. 2016).

It is essential to recognize the spine biomechanics, stresses, and pressures generated by neoplasmatic lesions to make the treatment most effective. Medical application of the FEM method is relatively rapid and simple. The only major constrain is that the method requires a proper bone modeling as a solid nonhomogeneous structure, from the standpoint of material properties. Studies on the method have worked away from the sheer recognition of forces acting on the spine and moved on to more complex examinations in vertebral pathologies (Matsuura et al. 2014; Unnikrishnan and Morgan 2011; Tschirhart et al. 2007; Whyne et al. 2003). The technique enables the picturing of forces acting on the spine based on CT pictures and may help design the optimum material and shape of surgical implants. It is useful when it comes to choose a proper method and technique of treatment, to plan a surgery and spine stabilization, with the benefit of better symptoms control and patients' quality of life. It is envisioned that the FEM in silico evaluation could be prospectively used in everyday orthopedic practice.

In conclusion, cancerous metastases to the lumbar spine lead to the formation of a pathological strain on the spine. Immobilization of the vertebral bodies around the spine fracture abolishes the strain; the effect being akin to that of the Jewett corset or surgical stabilization used to limit the motion of the spine. The finite element in silico method holds a strong potential to be helpful for a rapid designing of optimum shapes and choosing the proper material for surgical implants used in therapy.

Conflicts of Interest The authors declare no conflicts of interest in relation to this article.

References

Asdourian PL, Mardjetko S, Rauschning W, Jónsson H Jr, Hammerberg KW, Dewald RL (1990a) An evaluation of spinal deformity in metastatic breast cancer. J Spinal Disord 3:119–134

Asdourian PL, Weidenbaum M, De Wald RL, Hammerberg KW, Ramsey RG (1990b) The pattern of vertebral involvement in metastatic vertebral breast cancer. Clin Orthop Relat Res (250):164–170

Bartels RH, van der Linden YM, van der Graaf WT (2008) Spinal extradural metastasis: review of current treatment options. CA Cancer J Clin 58:245–259

Batson O (1942) The role of vertebral veins in metastatic processes. Ann Intern Med l:38–45

Bergmann G, Deuretzbacher G, Heller M, Greichen F, Rohlmann A, Strauss J, Duda DN (2001) Hip contact forces and gait patterns from routine activities. J Biomech 34(7):859–871

Ciftdemir M, Kaya M, Selcuk E, Yalniz E (2017) Tumors of the spine. World J Orthop 7:109–116

Coleman RE (2006) Clinical features of metastatic bone disease and risk of skeletal morbidity. Clin Cancer Res 12:6243–6249

David Roodman G, Silbermann R (2015) Mechanisms of osteolytic and osteoblastic skeletal lesions. Bonekey Rep 4:753

Denis F (1983) The three column spine and its significance in the classification of acute thoracolumbar spinal injuries. Spine (Phila Pa 1976) 8:817–831

Di Martino A, Caldaria A, De Vivo V, Denaro V (2016) Metastatic epidural spinal cord compression. Expert Rev Anticancer Ther 12:1–10

Ditunno JF, Young W, Donovan WH, Creasey G (1994) The international standards booklet for neurological and functional classification of spinal cord injury. American Spinal Injury Association Paraplegia 32:70–80

Dougall WC, Chaisson M (2006) The RANK/RANKL/OPG triad in cancer-induced bone diseases. Cancer Metastasis Rev 25:541–549

Ecker RD, Endo T, Wetjen NM, Krauss WE (2005) Diagnosis and treatment of vertebral column metastases. Mayo Clin Proc 80:1177–1186

Groblewska M, Mroczko B, Czygier M, Szmitkowski M (2008) Cytokines as markers of osteolysis in the diagnostics of patients with bone metastases. Postepy Hig Med Dosw 62:668–675

Guzik G (2016) The correspondence between magnetic resonance images and the clinical and intraoperative status of patients with spinal tumors. Curr Med Imaging Rev 12:149–155

Hauge EM, Qvesel D, Eriksen EF, Mosekilde L, Melsen F (2001) Cancellous bone remodeling occurs in specialized compartments lined by cells expressing osteoblastic markers. J Bone Miner Res 16:1575–1582

Jacobs WB, Perrin RG (2001) Evaluation and treatment of spinal metastases: an overview. Neurosurg Focus 11 (6):e10

Kato S, Hozumi T, Takeshita K, Kondo T, Goto T, Yamakawa K (2012) Neurological recovery after posterior decompression surgery for anterior dural compression in paralytic spinal metastasis. Arch Orthop Trauma Surg 132:765–771

Kimbung S, Loman N, Hedenfalk I (2015) Clinical and molecular complexity of breast cancer metastases. Semin Cancer Biol 35:85–95

Kintzelé L, Weber MA (2017) Imaging diagnostics in bone metastases. Radiologe 57:113–128

Kopaczewski B, Jankowski R, Nowak S (2009) Prognostic factors and therapeutic recommendations in patients with metastasis of the spine. Neuroskop 11:77–85

Liao J, Li X, Koh AJ, Berry JE, Thudi N, Rosol TJ, Pienta KJ, McCauley LK (2008) Tumor expressed PTHrP facilitates prostate cancer-induced osteoblastic lesions. Int J Cancer 123:2267–2278

Loblaw DA, Mitera G, Ford M, Laperriere NJ (2012) A 2011 updated systematic review and clinical practice guideline for the management of malignant extradural spinal cord compression. Int J Radiat Oncol Biol Phys 84(2):312–317

Lorkowski J, Mrzygłod M, Kotela A, Kotela I (2014) Application of rapid computer modeling in the analysis of the stabilization method in intraoperative femoral bone shaft fracture during revision hip arthroplasty- a case report. Pol Orthop Traumatol 79:138–144

Matsuura Y, Giambini H, Ogawa Y, Fang Z, Thoreson AR, Yaszemski MJ, Lu L, An KN (2014) Specimen-specific nonlinear finite element modeling to predict vertebrae fracture loads after vertebroplasty. Spine (Phila Pa 1976) 39:E1291–E1296

Mc Cormaek R, Taraikovie E, Gaines RW (1994) The load sharing classification of spine fractures. Spine 19:1741–1742

Paget S (1889) The distribution of secondary growths in cancer of the breast. Lancet 1:571–573

Pawlak WZ, Gabriel Wcisło G, Leśniewski-Kmak K, Korniluk J (2002) The metastatic bone disease as a first clinical symptom in cancer. Współczesna Onkologia 4:206–215

Phanphaisarn A, Patumanond J, Settakorn J, Chaiyawat P, Klangjorhor J, Pruksakorn D (2016) Prevalence and survival patterns of patients with bone metastasis from common cancers in Thailand. Asian Pac J Cancer Prev 17:4335–4340

Pokhrel D, Sood S, McClinton C, Shen X, Badkul R, Jiang H, Mallory M, Mitchell M, Wang F, Lominska C (2017) On the use of volumetric-modulated arc therapy for single-fraction thoracic vertebral metastases stereotactic body radiosurgery. Med Dosim 42:69–75

Sanderson RD, Yang Y (2008) Syndecan-1: a dynamic regulator of the myeloma microenvironment. Clin Exp Metastasis 25:149–159

Shiozawa Y, Pedersen EA, Havens AM, Jung Y, Mishra A, Joseph J, Kim JK, Patel LR, Ying C, Ziegler AM, Pienta MJ, Song J, Wang J, Loberg RD, Krebsbach PH, Pienta KJ, Taichman RS (2011) Human prostate cancer metastases target the hematopoietic stem cell niche to establish footholds in mouse bone marrow. J Clin Invest 121:1298–1312

Steinauer K, Gross MW, Huang DJ, Eppenberger-Castori S, Güth U (2014) Radiotherapy in patients with distant metastatic breast cancer. Radiat Oncol 9:126

Tokuhashi MH, Oda H, Oshima M, Ryu J (2005) A revised scoring system for preoperative evaluation of metastatic spine tumor prognosis. Spine (Phila Pa 1976) 30:2186–2191

Tomita K, Kawahara N, Kobayashi T, Yoshida A, Murakami H, Akamaru T (2001) Surgical strategy for spinal metastases. Spine 26:298–306

Tschirhart CE, Finkelstein JA, Whyne CM (2007) Biomechanics of vertebral level, geometry, and transcortical tumors in the metastatic spine. J Biomech Eng 40:46–54

Unnikrishnan GU, Morgan EF (2011) A new material mapping procedure for quantitative computed tomography-based, continuum finite element analyses of the vertebra. J Biomech Eng 133(7):071001

Whyne CM, Hu SS, Lotz JC (2003) Biomechanically derived guideline equations for burst fracture risk prediction in the metastatically involved spine. J Spinal Disord Tech 16:180–185

Wilson DA, Fusco DJ, Uschold TD, Spetzler RF, Chang SW (2012) Survival and functional outcome after surgical resection of intramedullary spinal cord metastases. World Neurosurg 77:370–374

Wirtz DC, Schiffers N, Pandorf T, Radermacher K, Weichert D, Forst R (2000) Critical evaluation of known bone material properties to realize anisotropic FE simulation of the proximal femur. J Biomech 33:1325–1330

Zhao C, Liang J, Yang Y, Yu M, Qu X (2017) The impact of glucagon-like peptide-1 on bone metabolism and its possible mechanisms. Front Endocrinol (Lausanne) 8:98

Adv Exp Med Biol - Clinical and Experimental Biomedicine (2018) 1: 41–52
https://doi.org/10.1007/5584_2018_188
© Springer International Publishing AG, part of Springer Nature 2018
Published online: 24 March 2018

Perspective on Broad-Acting Clinical Physiological Effects of Photobiomodulation

Steven Shanks and Gerry Leisman

Abstract

Research into photobiomodulation reveals beneficial effects of light therapy for a rapidly expanding list of medical conditions and illnesses. Although it has become more widely accepted by the mainstream medicine, the effects and mechanisms of action appear to be poorly understood. The therapeutic benefits of photobiomodulation using low-energy red lasers extend far beyond superficial applications, with a well-described physics allowing an understanding of how red lasers of certain optimum intensities may cross the cranium. We now have a model for explaining potential therapeusis for applications in functional neurology that include stroke, traumatic brain injury, and neurodegenerative conditions in addition to the currently approved functions in lipolysis, in onychomycosis treatment, and in pain management.

Keywords

Cold laser · Laser therapy · Lipolysis · Neurodegeneration · Onychomycosis · Pain ·

Photobiomodulation · Photobiostimulation · Stroke

1 Introduction

Photobiomodulation or low-level laser therapy, low-intensity laser therapy, low-power laser therapy, cold laser, soft-laser, or photobiostimulation has been studied for over 50 years (Anders et al. 2015; Mester and Jászsági-Nagy 1971), with new clinical applications of light therapies growing exponentially. Unfortunately, this has resulted in a growing confusion about the clinical effectiveness of this technology due to the diverse assortment of lasers and light-emitting diode (LED) devices employing different light wavelengths, each purportedly treating conditions from hair loss (Avci et al. 2014) to cancer and neurodegenerative disease (Santana-Blank et al. 2016). To alleviate some of this confusion, "photobiomodulation" (PBM) has recently been added to the Medical Subject Headings of the National Library of Medicine thesaurus for PubMed indexing. The

S. Shanks
Erchonia Corporation, Melbourne, FL, USA

G. Leisman (✉)
Faculty of Health Sciences, University of Haifa, Haifa, Israel

National Institute for Brain & Rehabilitation Sciences, Nazareth, Israel
e-mail: g.leisman@alumni.manchester.ac.uk

purpose of introducing a new term was to distinguish the PBM from other light-based devices that rely on thermal effects for some or all of their mechanisms of action.

Among the characteristics of PBM, including wavelength, power, and energy density, an emphasis has been placed on the ability of light to penetrate tissue (Hamblin 2017). Although PBM is well established as a beneficial treatment for numerous conditions, skepticism exists among those who think that light therapy, and particularly low-energy red lasers, can only be used to treat superficial conditions due to non-penetration of light energy in the body and bone. Far-red and near-infrared light waves do penetrate tissue to some extent. However, red lasers are capable of producing beneficial effects extending beyond treatment of superficial conditions (Grover et al. 2017; Maksimovich 2016; DeTaboada et al. 2006). It is apparent that the mode of therapeutic action of PBM is as complex as the pharmacokinetics of many medications. In the editorial in Photomedicine and Laser Surgery, Tiina Karu (2013) has asked a question: "Is it time to consider photobiomodulation as a drug equivalent?" She cited the use of PBM for treating a wide range of disorders including Parkinson's (Shaw et al. 2010) oral mucositis (Bjordal et al. 2011), peripheral nerve damage (Moges et al. 2011), ischemic tissue injury (Lapchak 2010), and other disorders (Santana-Blank et al. 2016).

The objective of this review is to provide perspective on the physiologically modifying properties of PBM, with potential application extending far beyond superficial applications including the evolving applications in functional neurology.

2   Photobiomodulation Effects

PBM employs non-ionizing light, including lasers, light-emitting diodes, or broadband light in the visible red (600–700 nm) and near-infrared (780–1100 nm) spectra to achieve therapeutic effects (de Freitas and Hamblin 2016). PBM is a nonthermal process beginning when a chromophore molecule is exposed to a suitable wavelength of light. Chromophores are responsible for the color associated with biological compounds such as hemoglobin, myoglobin, and cytochromes (Cotler et al. 2015). When a chromophore absorbs a photon of light, an electron transits to an excited state. A common target chromophore for PBM is the iron- and copper-containing enzyme cytochrome C oxidase in the mitochondrial respiratory chain, which absorbs light in the near-infrared spectrum (Avci et al. 2013; Karu and Afanasyeva 1995). The physiologic effects of PBM occur when photons dissociate the inhibitory signaling molecule of nitric oxide (NO), from cytochrome C oxidase, increasing the electron transport, mitochondrial membrane potentials, and the production of mitochondrial products such as ATP, NADH, RNA, and cellular respiration (Wang et al. 2016; de Freitas and Hamblin 2016).

Other effects include increases in the antioxidant enzyme activity, such as catalase and superoxide dismutase (Martins et al. 2016) or mitochondrial NO synthase, release of NO, also a potent vasodilator (Adamskaya et al. 2011), and the production of reactive oxygen species (ROS). ROS play an important role in cell signaling, cell cycle progression regulation, enzyme activation, and nucleic acid and protein synthesis (Holmström and Finkel 2014). ROS also activate transcription factors, leading to cellular proliferation, migration, and production of cytokines and growth factors (Farivar et al. 2014). Alternatively, light-sensitive ion channels can be opened to permit calcium ion entry into the cell.

Based on these complex characteristics, PBM possesses physiologically modifying properties associated with specific light characteristics, such as wavelength and irradiance, varied by exposure parameters, such as energy density, irradiation duration, and treatment frequency. Similarly to pharmacological agents, BPM displays a biphasic dose-response described by the Arndt-Schulz law (Huang et al. 2009). Increasing the PBM dose, based on exposure time or energy density, increases the response to a maximum effect, after which a further exposure increase results in decreased response or bioinhibition.

The biphasic dose-response of BPM has been demonstrated in in vitro (Solmaz et al. 2017; Hawkins and Abrahamse 2006) and in in vivo studies (de Lima et al. 2014; Hsieh et al. 2014; Rojas and Gonzalez-Lima 2013). These properties of BPM are sharply contrasted with infrared radiation (780–3000 nm), which heats tissues, resulting in physical changes.

PBM has been demonstrated to be beneficial for treating chronic joint disorders (Bjordal et al. 2003), musculoskeletal (Cotler et al. 2015), chronic low back (Huang et al. 2015) and neck pain (Gross et al. 2013; Chow et al. 2009), herniated disks (Takahashi et al. 2012), adhesive capsulitis (Ip and Fu 2015), and in wound healing (Aragona et al. 2017; Heidari et al. 2017; Yadav and Gupta 2017; Arany 2016; Kuffler 2016). PBM has even been determined to be promising for treating psychiatric and neurodegenerative disorders (Berman et al. 2017; Salehpour and Rasta 2017; Cassano et al. 2016; Hamblin 2016).

To date, clinical trials have demonstrated the therapeutic effects of PBM leading to the US-FDA clearance of devices for body sculpting (Roche et al. 2017; Thornfeldt et al. 2016; Suarez et al. 2014; McRae and Boris 2013; Jackson et al. 2012; Nestor et al. 2012; Jackson et al. 2009), cellulite (Jackson et al. 2013), pain (Roche et al. 2016), and, most recently, onychomycosis (Zang et al. 2017). There have been no reports of treatment-related adverse events.

3 Potential Clinical Applications

3.1 Lipolysis

Activation of cytochrome C oxidase triggers cellular events including an increase in ATP synthesis, with upregulation of cAMP and cytoplasmic lipase activation. Activated lipase breaks down intracellular triglycerides into fatty acids and glycerol (Karu and Afanasyeva 1995). An additional effect of cytochrome C oxidase activation is a transient formation of pores in the cell membrane of adipocytes, allowing the newly formed fatty acids and glycerol to pass into the extracellular space (Solarte et al. 2003). Electron microscopic images have demonstrated pore formation in the cell membranes of adipose cells exposed to low-level laser light. Upon entering the extracellular space, lipids released following PBM treatment are transported to lymph nodes, where lysosomal acid lipase hydrolyzes the released triglycerides to generate non-esterified free fatty acids (Neira et al. 2002). Alternatively, released lipids may be transported via the lymphatic system to the liver where they undergo normal fatty acid oxidation. Clinically, the use of PBM has been associated with decreased plasma triglycerides and cholesterol (Jackson et al. 2010; Rushdi 2010). Importantly, PBM does not result in necrosis, the endocrine function of adipose tissue is preserved (Poulos et al. 2010), and it prevents inflammatory effects of high-intensity focused ultrasound (Burks et al. 2011; Biermann et al. 2010) and cryolipolysis (Avram and Harry 2009).

3.2 Pain

A well-controlled trial has assessed PBM efficacy for treating chronic shoulder and neck pain and improving the upper body range of motion (ROM) (Roche et al. 2016). Participants received PBM using a diode laser emitting a divergent 635 nm (red) laser light (energy output of 1 mW) or sham treatment. After 48 h, 28 out of the 43 PBM-treated subjects (65.1%) demonstrated a greater than 30% improvement in pain scores vs. six (11.6%) in the sham-treated group.

The mechanism whereby PBM decreases pain is unknown (Hamblin 2017; Holanda et al. 2017). The analgesic and anti-inflammatory effects of PBM are associated with increased antioxidant glutathione and decreased expression of P2X3 receptor subunits in C- and Aδ-fiber afferent neurons (Janzadeh et al. 2016), significant reductions in cyclooxygenase-2 (COX-2) mRNA (Prianti et al. 2014), activation of endogenous opioids (Pereira et al. 2017), reductions in pro-inflammatory cytokines and glutamate, increases in endogenous analgesic prostatic acid phosphatase (Pires de Sousa et al. 2016), and the expression of bradykinin receptors (de Oliveira et al. 2017).

3.3   Onychomycosis

Zang et al. (2017) have reported on the use of PBM for treating onychomycosis. Fifty affected toenails were treated for 12 min weekly for 2 or 4 weeks with a 635/405 nm dual-diode laser device. Most treated toenails (67%) achieved individual treatment success. The extent of clearing at the nail baseline increased by a mean of 4.8 ± 5.2(SD) mm after 6 months. Nearly all treated toenails (89%) demonstrated an increase in the area of clear nails over the 6-month study period. Based on the safety and efficacy, this nonthermal laser PBM device received FDA 510 (k) market clearance for a temporary nail clearing in patients with onychomycosis caused by dermatophytes or yeasts.

PBM also exhibits antimicrobial effects on biofilms formed by *Streptococcus mutans* and *Candida albicans* (Basso et al. 2011) and activity against *C. albicans* cultures (Maver-Biscanin and Mravak-Stipetic 2005). It also is effective against oral *C. albicans* infections in mice and humans in vivo (Seyedmousavi et al. 2014; Scwingel et al. 2012) and against the fungus *Paracoccidioides brasiliensis* in both in vitro and in vivo (Burger et al. 2015).

The mechanisms of antifungal effect of PBM are unclear. Exposure of cytochrome C oxidase in the mitochondrial respiratory chain to suitable wavelength of light results in increased production of mitochondrial ATP, NADH, and RNA and increased cellular respiration (de Freitas and Hamblin 2016; Wang et al. 2015; Chung et al. 2012). Consistent with these effects, neutrophils from PBM-treated mice become more metabolically active and have higher fungicidal activity (Burger et al. 2015). When stimulated by PBM in vitro, human neutrophils show a greater production of ROS and increased fungicidal capacity against *C. albicans* (Cerdeira et al. 2016).

In contrast, thermal lasers used for treating onychomycosis emit light in the near-infrared spectrum (780–3000 nm) and exert their effect by simply heating the target tissue (Nenoff et al. 2014). In vitro studies have demonstrated a thermal killing effect on fungal mycelia when treatment temperatures exceed 50 °C (Carney et al. 2013; Paasch et al. 2013). The mean peak temperatures associated with an 808 nm laser is 74.1–112.4 °C and that of a 980 nm laser is 45.8–53.5 °C (Paasch et al. 2014). Consequently, the use of thermal lasers has been associated with moderate pain or burning sensations (Helou et al. 2016; Moon et al. 2014; Noguchi et al. 2013) and darkening under the nail (Lu et al. 2016). Discomfort is minimized by using a pulsed-wave laser instead of continuous-wave laser (Anderson and Parrish 1983).

4   Functional Neurological Applications

4.1   Photobiomodulation and the Cranial Vault

A significant literature exists on the ability of PBM to penetrate the skull in both diagnostic and therapeutic applications. Low-energy laser passes the skull and a therapeutic effect likely exists. Low-energy laser systems employ the so-called quantum optical induced transparency (QIT) effect (Weis et al. 2010; Harris et al. 1990). Quantum interference in the amplitude of optical transitions in atomic media can lead to strong modification in optical properties. This effect, electromagnetically induced transparency, controls optical properties of dense media and can enhance transparency contrast by a factor of five (Scherman et al. 2012). Therefore, the skull, spine, or joints can be penetrated even with moderate intensity light. Due to the QIT effect, the radiation should reach deep tissue layers in muscles, connective tissue, and even bone, enabling noninvasive transcranial treatments for neurodegenerative diseases, stroke, or traumatic brain injury (TBI) (Tedford et al. 2015; Stemer et al. 2010; Oron et al. 2006). Litscher and Litscher (2013) have reported that laser emission in the yellow band of the light spectrum could penetrate the human cranium making PBM a promising option for treating neurodegeneration or stroke, TBI, and other neurologically based conditions. A central constraint controlling the depth of penetration of laser light past the cranium is wavelength. Both absorption and scattering coefficients of living tissue are greater at lesser wavelengths with near-infrared light penetrating more deeply than red light. It also is argued that pulsed-wave lasers penetrate more deeply into tissue than continuous-wave lasers with the same average energy power.

Let us suppose that at a wavelength of 810 nm, the depth of tissue at which the laser intensity is reduced to 10% of its value at the skin surface is 1 cm. A laser, with a power density (irradiance) of 100 mW/cm^2 at the skin, would have a power density of 10 mW/cm^2 1 cm below the skin and of 1 mW/cm^2 2 cm below the skin. Now, suppose that a threshold power density, a minimum number of photons *per* unit area *per* unit time, at the target tissue is necessary to have a biological effect and that this value is 10 mW/cm^2. The effective penetration depth of continuous-wave laser may be, say, 1 cm. By contrast, let us consider a pulsed-wave laser with a 10 ms pulse duration and frequency of 1 Hz (direct current (DC) =1 Hz × 0.010 s = 0.010) and the same average energy power. The peak power and power densities are now 100 times higher (peak power = average power/DC = average power × 100). With a peak power density of 10 W/cm^2 at the skin, the tissue depth, at which this peak power density attenuates to the threshold level of 10 mW/cm^2, is now 3 cm rather than 1 cm in the continuous-wave mode. However, it should be taken into consideration that the laser is on only for 1% of the time, so that the total fluence delivered to 3 cm depth in the pulse-waved mode is 100 times less than that delivered to 1 cm depth in the continuous mode.

The use of PBM to treat diseases, disorders, and injuries of the brain requires a detailed understanding of the nature of light propagation through tissues including the scalp, skull, meninges, and brain. Tedford et al. (2015) have investigated light penetration gradients in the human cadaver brains using a transcranial laser system with a wavelength of 808 nm and wavelength dependence of light scatter and absorbance in intraparenchymal brain tissue using 660, 808, and 940 nm. Those authors demonstrate that the 808 nm wavelength light has a superior brain tissue penetration.

4.2 Photobiomodulation (PBM) Application for Neurodegenerative Diseases

Peptide fibrillization is associated with the creation of amyloid fibers in Parkinson's and Alzheimer's diseases (Luo et al. 2000). Hanczyc et al. (2012) have shown that amyloids demonstrate robust nonlinear optical absorption that is not extant in native non-fibrillized proteins. Pérez-Moreno et al. (2008) have found that lysozyme β-amyloids, insulin, and α-synuclein display two-, three-, or multiphoton absorption processes, depending on the wavelength of light. Olesiak-Banska et al. (2012) have suggested that a heightened multiphoton absorption is the outcome of a mechanism comprising dipolar through-space pairing between excited states of aromatic amino acids compacted in fibrous assemblies (Figs. 1 and 2).

Moro et al. (2014) have investigated neuroprotection offered by PBM in the mouse model. Mice were treated with PBM 20 times over 4 weeks and Alzheimer's disease-related histochemical markers were examined in the neocortex and hippocampus. Those authors show that PBM treatment decreases neurofibrillary tangles, hyperphosphorylated tau protein, and oxidative stress markers, as well as restores cytochrome C oxidase, which suggests that it may potentially be

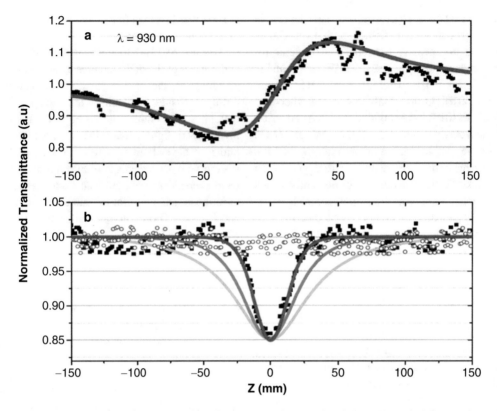

Fig. 1 (**a**) Z-scan example measurement of nonlinear refractive index and (**b**) nonlinear absorption coefficient of amyloid sample. Femtosecond Z-scan using 100 GW/cm^2 irradiation at 930 nm with closed (**a**) and open openings. (**b**) Amyloid fibrils (squares) are compared to native protein (circles). Theoretical fits are shown for two-photon absorption (2PA, light-gray solid curve), three-photon absorption (3PA, dark-gray solid curve), and five-photon absorption (5PA, red solid curve) (With permission from Olesiak-Banska et al. 2012)

an effective, minimally invasive intervention for progressive cerebral degeneration.

As downregulation of hippocampal brain-derived neurotrophic factor (BDNF), necessary for neuronal survival and dendritic sprouting, occurs in early Alzheimer's disease, BDNF upregulation may provide a mechanism to save dendritic atrophy in the disease course. Meng et al. (2013) have found that PBM reverses a decline in dendritic atrophy and in the number of neurons by means of BDNF upregulation. The PBM, through modulation of the transcription factor cAMP response element-binding protein (CREB), increases BDNF, mRNA, protein expression, as well as dendrite growth (dendritic spine length, density, and branching in hippocampal neurons).

4.3 Stroke

In an early study with transcranial-PBM (T-BPM) for stroke treatment in the rat model, Zhang et al. (1997) have obtained significant results. T-BPM at 808 nm, a previously alluded to important wavelength, significantly improves recovery 3 weeks after ischemic stroke. The regimen included one treatment on the contra-lesional side (power density: 7.5 mW/cm^2) 24 h post-stroke, with a good result in terms of neurogenesis stimulation.

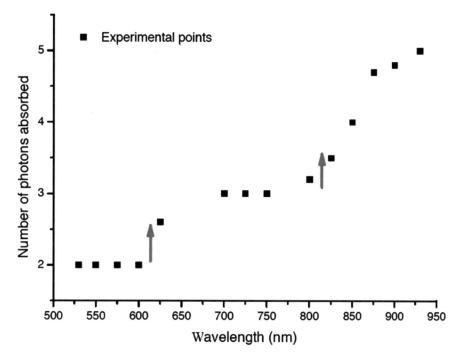

Fig. 2 Wavelength dependence of exponent n (the apparent number of photons taking part in the nonlinear absorption process) defined from dI/dz = -kIn with z denoting propagation distance I denoting light intensity. Steps marked with vertical arrows: below 625 nm, two-photon absorption and below 850 nm, three-photon absorption (From Olesiak-Banska et al. 2012 with permission)

A study in embolized rabbits has shown a direct relationship between cortical fluence (energy density, J/cm^2) and cortical ATP (Lapchak and DeTaboada 2010). Five minutes following embolization, rabbits received a 2-min PBM using 808 nm laser on the skin surface. Continuous-wave (CW) PBM (7.5 mW/cm^2, 0.9 J/cm^2) produced a 41% increase in cortical ATP. A 100 Hz pulse wave (PW) PBM (37.5 mW/cm^2, 4.5 J/cm^2) produced a 157% increase in cortical ATP. The authors suggest that even a greater improvement potential might be achievable by treatment length and mode (PW, at approximately 100 Hz) optimization.

T-BPM has been shown to significantly improve outcome in human stroke patients, when administered 18 h post-stroke, over the entire head surface, regardless of stroke localization (Lampl et al. 2007). Significant improvements are observed in the moderate and moderate–severe patients, but not in severe stroke patients (Zivin et al. 2009).

4.4 Traumatic Brain Injury (TBI)

Mild traumatic brain injury (mTBI) patients demonstrate cognitive and memory difficulties for at least 6 months or later after an episode. There is a significant requirement for effectual methods to foster cognitive recovery (Cicerone et al. 2006). mTBI from single and multiple events has been the most recurrent injury experienced by military personnel in the Operation Enduring Freedom and Operation Iraqi Freedom (Hoge et al. 2008). Diffuse axonal injury is frequently evidenced in frontotemporal and anterior-corona-radiata regions (Niogi et al. 2008; Taber et al. 2006). Cognitive impairments result from tissue injury in the prefrontal and frontal cortical regions and anterior cingulate gyrus within the frontal lobes. PBM has been applied in animal TBI models (Oron et al. 2007). Mice were subjected to closed head injury. Five days later, motor behavior was significantly better in the PBM-treated group. Twenty-eight days post-injury, the mean lesion size in the laser-treated group was significantly smaller than in controls. The PBM has also had positive effects in other studies (Wu et al. 2012; Moreira et al. 2009). Naeser et al. (2011) have reported in humans that chronic mTBI cases demonstrate improved cognitive function after PBM that is associated with increased ATP. In a recent study on major depression treatment, Schiffer et al. (2009) have shown that PBM increases regional cerebral blood flow in the contra-lesional frontal lobes (power density: 7.5 mW/cm^2). It also increases cellular respiration and oxygenation of hypoxic cells (Mi et al. 2004a and b).

5   Conclusions

Research into photobiomodulation is revealing the beneficial effects of light therapy for a rapidly expanding list of medical conditions and illnesses. As it becomes more widely accepted by the mainstream medicine, its mechanism of action and the biophysics appear to be poorly understood. Importantly, therapeutic benefits of photobiomodulation, using red lasers, extend far beyond superficial applications.

Acknowledgments The authors acknowledge the editorial assistance of Dr. Carl S. Hornfeldt.

Competing Interests S.S. is president of the Erchonia Corporation in Melbourne, Florida; G.L. has no fiduciary relationship with any manufacturer or marketer of therapeutic laser equipment.

References

Adamskaya N, Dungel P, Mittermayr R, Hartinger J, Feichtinger G, Wassermann K, Redl H, van Griensven M (2011) Light therapy by blue LED improves wound healing in an excision model in rats. Injury 42:917–921

Anders JJ, Lanzafame RJ, Arany PR (2015) Low-level light/laser therapy *versus* photobiomodulation therapy. Photomed Laser Surg 33:183–184

Anderson RR, Parrish JA (1983) Selective photothermolysis: precise microsurgery by selective absorption of pulsed radiation. Science 220:524–527

Aragona SE, Grassi FR, Nardi G, Lotti J, Mereghetti G, Canavesi E, Equizi E, Puccio AM, Lotti T (2017) Photobiomodulation with polarized light in the treatment of cutaneous and mucosal ulcerative lesions. J Biol Regul Homeost Agents 31(2 Suppl 2):213–218

Arany PR (2016) Craniofacial wound healing with photobiomodulation therapy: new insights and current challenges. J Dent Res 95:977–984

Avci P, Gupta A, Sadasivam M, Vecchio D, Pam Z, Pam N, Hamblin MR (2013) Low-level laser (light) therapy (LLLT) in skin: stimulating, healing, restoring. Semin Cutan Med Surg 32:41–52

Avci P, Gupta GK, Clark J, Wikonkal N, Hamblin MR (2014) Low-level laser (light) therapy (LLLT) for treatment of hair loss. Lasers Surg Med 46:144–151

Avram MM, Harry RS (2009) Cryolipolysis for subcutaneous fat layer reduction. Lasers Surg Med 41:703–708

Basso FG, Oliveira CF, Fontana A, Kurachi C, Bagnato VS, Spolidório DM, Hebling J, de Souza Costa CA (2011) In vitro effect of low-level laser therapy on typical oral microbial biofilms. Braz Dent J 22:502–510

Berman MH, Halper JP, Nichols TW, Jarrett H, Lundy A, Huang JH (2017) Photobiomodulation with near infrared light helmet in a pilot, placebo-controlled clinical trial in dementia patients testing memory and cognition. J Neurol Neurosci 8:176

Biermann K, Montironi R, Lopez-Beltran A, Zhang S, Cheng L (2010) Histopathological findings after treatment of prostate cancer using high-intensity focused ultrasound (HIFU). Prostate 70:1196–1200

Bjordal JM, Couppé C, Chow RT, Tunér J, Ljunggren EA (2003) A systematic review of low level laser therapy with location-specific doses for pain from chronic joint disorders. Aust J Physiother 49:107–116

Bjordal JM, Bensadoun RJ, Tunèr J, Frigo L, Gjerde K, Lopes-Martins RA (2011) A systematic review with meta-analysis of the effect of low-level laser therapy (LLLT) in cancer therapy-induced oral mucositis. Support Care Cancer 19:1069–1077

Burger E, Mendes AC, Bani GM, Brigagão MR, Santos GB, Malaquias LC, Chavasco JK, Verinaud LM, de Camargo ZP, Hamblin MR, Sperandio FF (2015) Low-level laser therapy to the mouse femur enhances the fungicidal response of neutrophils against *Paracoccidioides brasiliensis*. PLoS Negl Trop Dis 9 (2):e0003541

Burks SR, Ziadloo A, Hancock HA, Chaudhry A, Dean DD, Lewis BK, Frenkel V, Frank JA (2011) Investigation of cellular and molecular responses to pulsed focused ultrasound in a mouse model. PLoS One 6 (9):e24730

Carney C, Cantrell W, Warner J, Elewski B (2013) Treatment of onychomycosis using a submillisecond 1064-nm neodymium:yttrium-aluminum-garnet laser. J Am Acad Dermatol 69:578–582

Cassano P, Petrie SR, Hamblin MR, Henderson TA, Iosifescu DV (2016) Review of transcranial photobiomodulation for major depressive disorder: targeting brain metabolism, inflammation, oxidative stress, and neurogenesis. Neurophotonics 3:031404

Cerdeira CD, Lima Brigagão MR, Carli ML, de Souza Ferreira C, de Oliveira Isac Moraes G, Hadad H, Costa Hanemann JA, Hamblin MR, Sperandio FF (2016) Low-level laser therapy stimulates the oxidative burst in human neutrophils and increases their fungicidal capacity. J Biophotonics 9:1180–1188

Chow RT, Johnson MI, Lopes-Martins RA, Bjordal JM (2009) Efficacy of low-level laser therapy in the management of neck pain: a systemic review and meta-analysis of randomized placebo or active-treatment controlled trials. Lancet 374:1897–1908

Chung H, Dai T, Sharma SK, Huang YY, Carroll JD, Hamblin MR (2012) The nuts and bolts of low-level laser (light) therapy. Ann Biomed Eng 40:516–533

Cicerone K, Levin H, Malec J, Stuss D, Whyte J (2006) Cognitive rehabilitation interventions for executive function: moving from bench to bedside in patients withtraumatic brain injury. J Cogn Neurosci 18:1212–1222

Cotler HB, Chow RT, Hamblin MR, Carroll J (2015) The use of low level laser therapy (LLLT) for musculoskeletal pain. MOJ Orthop Rheumatol 2(5):pii 00068

de Freitas LF, Hamblin MR (2016) Proposed mechanisms of photobiomodulation or low-level light therapy. IEEE J Sel Top Quantum Electron 22:7000417

de Lima FM, Aimbire F, Miranda H, Vieira RP, deOliveira AP, Albertini R (2014) Low-level laser therapy attenuates the myeloperoxidase activity and inflammatory mediator generation in lung inflammation induced by gut ischemia and reperfusion: a dose-response study. J Lasers Med Sci 5:63–70

de Oliveira VL, Silva JA Jr, Serra AJ, Pallota RC, da Silva EA, de Farias Marques AC, Feliciano RD, Marcos RL, Leal-Junior EC, de Carvalho PT (2017) Photobiomodulation therapy in the modulation of inflammatory mediators and bradykinin receptors in an experimental model of acute osteoarthritis. Lasers Med Sci 32:87–94

DeTaboada L, Ilic S, Leichliter-Martha S, Oron U, Oron A, Streeter J (2006) Transcranial application of low-energy laser irradiation improves neurological deficits in rats following acute stroke. Lasers Surg Med 38:70–73

Farivar S, Malekshahabi T, Shiari R (2014) Biological effects of low level laser therapy. J Lasers Med Sci 5:58–62

Gross AR, Dziengo S, Boers O, Goldsmith CH, Graham N, Lilge L, Burnie S, White R (2013) Low level laser therapy (LLLT) for neck pain: a systematic review and meta-regression. Open Orthop J 7:396–419

Grover F Jr, Weston J, Weston M (2017) Acute effects of near infrared light therapy on brain state in healthy subjects as quantified by qEEG measures. Photomed Laser Surg 35:136–141

Hamblin MR (2016) Shining light on the head: photobiomodulation for brain disorders. BBA Clin 6:113–124

Hamblin MR (2017) Mechanisms and applications of the anti-inflammatory effects of photobiomodulation. AIMS Biophys 4:337–361

Hanczyc P, Norden B, Samoc M (2012) Two-photon absorption of metal-organic DNA-probes. Dalton Trans 41:3123–3125

Harris SE, Field JE, Imamoglu A (1990) Nonlinear optical processes using electromagnetically induced transparency. Phys Rev Lett 64:1107–1110

Hawkins DH, Abrahamse H (2006) The role of laser fluence in cell viability, proliferation, and membrane integrity of wounded human skin fibroblasts following helium-neon laser irradiation. Lasers Surg Med 38:74–83

Heidari M, Paknejad M, Jamali R, Nokhbatolfoghahaei H, Fekrazad R, Moslemi N (2017) Effect of laser photobiomodulation on wound healing and postoperative pain following free gingival graft: a split-mouth triple-blind randomized controlled clinical trial. J Photochem Photobiol B 172:109–114

Helou J, Maatouk I, Hajjar MA, Moutran R (2016) Evaluation of Nd:YAG laser device efficacy on onychomycosis: a case series of 30 patients. Mycoses 59:7–11

Hoge CW, McGurk D, Thomas JL, Cox AL, Engel CC, Castro CA (2008) Mild traumatic brain injury in U.S. soldiers returning from Iraq. N Engl J Med 358:453–463

Holanda VM, Chavantes MC, Wu X, Anders JJ (2017) The mechanistic basis for photobiomodulation therapy of neuropathic pain by near infrared laser light. Lasers Surg Med 49:516–524

Holmström KM, Finkel T (2014) Cellular mechanisms and physiological consequences of redox-dependent signalling. Rev Mol Cell Biol 15:411–421

Hsieh YL, Cheng YJ, Huang FC, Yang CC (2014) The fluence effects of low-level laser therapy on inflammation, fibroblast-like synoviocytes, and synovial apoptosis in rats with adjuvant-induced arthritis. Photomed Laser Surg 32:669–677

Huang YY, Chen AC, Carroll JD, Hamblin MR (2009) Biphasic dose response in low level light therapy. Dose Response 7:358–383

Huang Z, Ma J, Chen J, Shen B, Pei F, Kraus VB (2015) The effectiveness of low-level laser therapy for non-specific chronic low back pain, a systematic review and meta-analysis. Arthritis Res Ther 17:360

Ip D, Fu NY (2015) Two-year follow-up of low-level laser therapy for elderly with painful adhesive capsulitis of the shoulder. J Pain Res 8:247–252

Jackson RF, Dedo DD, Roche GC, Turok DI, Maloney RJ (2009) Low-level laser therapy as a non-invasive approach for body contouring, a randomized, controlled study. Lasers Surg Med 41:799–809

Jackson RF, Roche GC, Wisler K (2010) Reduction in cholesterol and triglyceride serum levels following low-level laser irradiation: a noncontrolled, nonrandomized pilot study. Am J Cosmet Surg 27:177–184

Jackson RF, Stern FA, Neira R, Ortiz-Neira CL, Maloney J (2012) Application of low-level laser therapy for non-invasive body contouring. Lasers Surg Med 44:211–217

Jackson RF, Roche GC, Shanks SC (2013) A double-blind, placebo-controlled randomized trial evaluating the ability of low-level laser therapy to improve the appearance of cellulite. Lasers Surg Med 45:141–147

Janzadeh A, Nasirinezhad F, Masoumipoor M, Jameie SB, Hayat P (2016) Photobiomodulation therapy reduces apoptotic factors and increases glutathione levels in a neuropathic pain model. Lasers Med Sci 31:1863–1869

Karu T (2013) Is it time to consider photobiomodulation as a drug equivalent? Photomed Laser Surg 31:189–191

Karu TI, Afanasyeva NI (1995) Cytochrome C oxidase as primary photoacceptor for cultured cells in visible and near IR regions. Doklady Akad Nauk (Moscow) 342:693–695

Kuffler DP (2016) Photobiomodulation in promoting wound healing: a review. Regen Med 11:107–122

Lampl Y, Zivin JA, Fisher M, Lew R, Welin L, Dahlof B, Borenstein P, Andersson B, Perez J, Caparo C, Ilic S, Oron U (2007) Infrared laser therapy for ischemic stroke, a new treatment strategy: results of the NeuroThera effectiveness and safety Trial-1 (NEST-1). Stroke 38:1843–1849

Lapchak PA (2010) Taking a light approach to treating acute ischemic stroke patients: transcranial near-infrared laser therapy translational science. Ann Med 42:576–586

Lapchak PA, DeTaboada L (2010) Transcranial near infrared laser treatment (NILT) increases cortical adenosine- 5′-triphosphate (ATP) content following embolic strokes in rabbits. Brain Res 1306:100–105

Litscher D, Litscher G (2013) Laser therapy and stroke: quantification of methodological requirements in consideration of yellow laser. Int J Photoen 2013:575798. https://doi.org/10.1155/2013/575798

Lu S, Zhang J, Liang Y, Li X, Cai W, Xi L (2016) The efficacy and prognostic factors for long pulse neodymium: yttrium-aluminum-garnet laser treatment on onychomycosis: a pilot study. Ann Dermatol 28:406–408

Luo Y, Norman P, Macak P, Agren H (2000) Solvent-induced two-photon absorption of a push-pull molecule. J Phys Chem A 104:4718–4722

Maksimovich IV (2016) Brain disorders and therapy. Brain 5:1000209

Martins DF, Turnes BL, Cidral-Filho FJ, Bobinski F, Rosas RF, Danielski LG, Petronilho F, Santos AR (2016) Light-emitting diode therapy reduces persistent inflammatory pain: role of interleukin 10 and antioxidant enzymes. Neuroscience 324:485–495

Maver-Biscanin M, Mravak-Stipetic VJ (2005) Effect of low-level laser therapy on *Candida albicans* growth in patients with denture stomatitis. Photomed Laser Surg 23:328–332

McRae E, Boris J (2013) Independent evaluation of low-level laser therapy at 635 nm for non-invasive body contouring of the waist, hips, and thighs. Lasers Surg Med 45:1–7

Meng C, He Z, Xing D (2013) Low-level laser therapy rescues dendrite atrophy via upregulating BDNF expression: implications for Alzheimer's disease. J Neurosci 33:13505–13517

Mester E, Jászsági-Nagy E (1971) Biological effects of laser radiation. Radiobiol Radiother (Berl) 12:377–385

Mi XQ, Chen JY, Liang ZJ, Zhou LW (2004a) *In vitro* effects of helium–neon laser irradiation on human blood: blood viscosity and deformability of erythrocytes. Photomed Laser Surg 22:477–482

Mi XQ, Chen JY, Cen Y, Liang ZJ, Zhou LW (2004b) A comparative study of 632.8 and 532 nm laser irradiation on some rheological factors in human blood *in vitro*. J Photochem Photobiol B 74:7–12

Moges H, Wu X, McCoy J, Vasconcelos OM, Bryant H, Grunberg NE, Anders JJ (2011) Effect of 810 nm light on nerve regeneration after autograft repair of severely injured rat median nerve. Lasers Surg Med 43:901–906

Moon SH, Hur H, Oh YJ, Choi KH, Kim JE, Ko JY, Ro YS (2014) Treatment of onychomycosis with a 1,064-nm long-pulsed Nd:YAG laser. J Cosmet Laser Ther 16:165–167

Moreira MS, Velasco IT, Ferreira LS, Ariga SK, Barbeiro DF, Meneguzzo DT, Abatepaulo F, Marques MM (2009) Effect of phototherapy with low intensity laser on local and systemic immunomodulation following focal brain damage inrat. J Photochem Photobiol B 97:145–151

Moro C, Massri NE, Torres N, Ratel D, De Jaeger X, Chabrol C, Johnstone D (2014) Photobiomodulation inside the brain: a novel method of applying near-infrared light intracranially and its impact on dopaminergic cell survival in MPTP-treated mice. J Neurosurg 120:670–683

Naeser MA, Saltmarche A, Krengel MH, Hamblin MR, Knight JA (2011) Improved cognitive functionaftertranscranial, light-emitting diode treatments in chronic, traumatic brain injury: two case reports. Photomed Laser Surg 29:351–358

Neira R, Arroyave J, Ramirez H, Ortiz CL, Solarte E, Sequeda F, Gutierrez MI (2002) Fat liquefaction: effect of low-level laser energy on adipose tissue. Plast Reconstr Surg 110:912–922

Nenoff P, Grunewald S, Paasch U (2014) Laser therapy of onychomycosis. J Dtsch Dermatol Ges 12:33–38

Nestor MS, Zarraga MB, Park H (2012) Effect of 635nm low-level laser therapy on upper arm circumference reduction: a double-blind, randomized, sham-controlled trial. J Clin Aesthet Dermatol 5:42–48

Niogi SN, Mukherjee P, Ghajar J, Johnson C, Kolster RA, Sarkar R, Lee H, Meeker M, Zimmerman RD, Manley GT, McCandliss BD (2008) Extent of microstructural white matter injury in postconcussive syndrome correlates with impaired cognitive reaction time: a 3T diffusion tensor imaging study of mild traumatic brain injury. AJNR Am J Neuroradiol 29:967–973

Noguchi H, Miyata K, Sugita T, Hiruma M, Hiruma M (2013) Treatment of onychomycosis using a 1064nm Nd:YAG laser. Med Mycol J 54:333–339

Olesiak-Banska J, Hanczyc P, Matczyszyn K, Norden B, Samoc M (2012) Nonlinear absorption spectra of ethidium and ethidium homodimer. Chem Phys 404:33–35

Oron A, Oron U, Chen J, Eilam A, Zhang C, Sadeh M, Lampl Y, Streeter J, DeTaboada L, Chopp M (2006) Low-level laser therapy applied transcranially to rats after induction of stroke significantly reduces long-term neurological deficits. Stroke 37:2620–2624

Oron A, Oron U, Streeter J, de Taboada L, Alexandrovich A, Trembovler V, Shohami E (2007) Low-level laser therapy applied transcranially to mice following traumatic brain injury significantly reduces long-term neurological deficits. J Neurotrauma 24:651–656

Paasch U, Mock A, Grunewald S, Bodendorf MO, Kendler M, Seitz AT, Simon JC, Nenoff P (2013) Antifungal efficacy of lasers against dermatophytes and yeasts in vitro. Int J Hyperth 29:544–550

Paasch U, Nenoff P, Seitz AT (2014) Heat profiles of laser-irradiated nails. J Biomed Opt 19(1):18001

Pereira FC, Parisi JR, Maglioni CB, Machado GB, Barragán-Iglesias P, Silva JRT, Silva ML (2017) Antinociceptive effects of low-level laser therapy at 3 and 8 J/cm^2 in a rat model of postoperative pain: possible role of endogenous opioids. Lasers Surg Med 49:844–851

Pérez-Moreno J, Clays K, Kuzyk MG (2008) A new dipole-free sum-over-states expression for the second hyperpolarizability. J Chem Phys 128:084109

Pires de Sousa MV, Ferraresi C, Kawakubo M, Kaippert B, Yoshimura EM, Hamblin MR (2016) Transcranial low-level laser therapy (810 nm) temporarily inhibits peripheral nociception: photoneuromodulation of glutamate receptors, prostatic acid phophatase, and adenosine triphosphate. Neurophotonics 3:015003

Poulos SP, Hausman DB, Hausman GJ (2010) The development and endocrine functions of adipose tissue. Mol Cell Endocrinol 323:20–34

Prianti AC Jr, Silva JA Jr, Dos Santos RF, Rosseti IB, Costa MS (2014) Low-level laser therapy (LLLT) reduces the COX-2 mRNA expression in both subplantar and total brain tissues in the model of peripheral inflammation induced by administration of carrageenan. Lasers Med Sci 29:397–1403

Roche GC, Murphy DJ, Berry TS, Shanks S (2016) Low-level laser therapy for the treatment of chronic neck and shoulder pain. Funct Neurol Rehabil Ergon 6:97–104

Roche GC, Shanks S, Jackson RF, Holsey LJ (2017) Low-level laser therapy for reducing the hip, waist, and upper abdomen circumference of individuals with obesity. Photomed Laser Surg 35:142–149

Rojas JC, Gonzalez-Lima F (2013) Neurological and psychological applications of transcranial lasers and LEDs. Biochem Pharmacol 86:447–457

Rushdi TA (2010) Effect of low-level laser therapy on cholesterol and triglyceride serum levels in ICU patients: a controlled, randomized study. EJCTA 4:95–99

Salehpour F, Rasta SH (2017) The potential of transcranial photobiomodulation therapy for treatment of major depressive disorder. Rev Neurosci 28:441–453

Santana-Blank L, Rodríguez-Santana E, Santana-Rodríguez KE, Reyes H (2016) Quantum leap in photobiomodulation therapy ushers in a new generation of light-based treatments for cancer and other complex diseases: perspective and mini-review. Photomed Laser Surg 34:93–101

Scherman M, Mishina OS, Lombardi P, Giacobino E, Laurat J (2012) Enhancing electromagnetically-induced transparency in a multilevel broadened medium. Opt Express 20:4346–4351

Schiffer F, Johnston AL, Ravichandran C, Polcari A, Teicher MH, Webb RH, Hamblin MR (2009) Psychological benefits 2 and 4 weeks after a single treatment with NIR light to the forehead: a pilot study of 10 patients with major depression and anxiety. Behav Brain Funct 5:46

Scwingel AR, Barcessat AR, Núñez SC, Ribeiro MS (2012) Antimicrobial photodynamic therapy in the treatment of oral candidiasis in HIV-infected patients. Photomed Laser Surg 30:429–432

Seyedmousavi S, Hashemi SJ, Rezaie S, Fateh M, Djavid GE, Zibafar E, Morsali F, Zand N, Alinaghizadeh M, Ataie-Fashtami L (2014) Effects of low-level laser irradiation on the pathogenicity of Candida albicans: in vitro and in vivo study. Photomed Laser Surg 32:322–329

Shaw VE, Spana S, Ashkan K, Benabid AL, Stone J, Baker GE, Mitrofanis J (2010) Neuroprotection in midbrain dopaminergic cells in MPTP-treated mice after near-infrared light treatment. J Comp Neurol 518:25–40

Solarte E, Isaza C, Criollo W, Rebolledo AF, Arroyave J, Ramirez H, Neira R (2003) *In vitro* effects of 635 nm low intensity diode laser irradiation on the fat distribution of one adipose cell. Proceedings SPIE, 19th Congress of the International Commission for Optics: Optics for the Quality of Life 4829:96l. https://doi.org/10.1117/12.527513

Solmaz H, Ulgen Y, Gulsoy M (2017) Photobiomodulation of wound healing via visible and infrared laser irradiation. Lasers Med Sci 32:903–910

Stemer AB, Huisa BN, Zivin JA (2010) The evolution of transcranial laser therapy for acute ischemic stroke, including a pooled analysis of NEST-1 and NEST-2. Curr Cardiol Rep 12:29–33

Suarez DP, Roche GC, Jackson RF (2014) A double-blind, sham-controlled study demonstrating the effectiveness of low-level laser therapy using a 532-nm green diode for contouring the waist, hips, and thighs. Am J Cosmet Surg 31:34–41

Taber KH, Warden DL, Hurley RA (2006) Blastrelated traumatic brain injury: what is known? J Neuropsychiatry Clin Neurosci 18:141–145

Takahashi H, Okuni I, Ushigome N, Harada T, Tsuruoka H, Ohshiro T, Sekiguchi M, Musya Y (2012) Low level laser therapy for patients with cervical disk hernia. Laser Ther 21:193–197

Tedford CE, DeLapp S, Jacques S, Anders J (2015) Quantitative analysis of transcranial and intraparenchymal light penetration in human cadaver brain tissue. Lasers Surg Med 47:312–322

Thornfeldt CR, Thaxton PM, Hornfeldt CS (2016) A six-week low-level laser therapy protocol is effective for reducing waist, hip, thigh, and upper abdomen circumference. J Clin Aesthet Dermatol 9:31–35

Wang L, Hu L, Grygorczyk R, Shen X, Schwarz W (2015) Modulation of extracellular ATP content of mast cells and DRG neurons by irradiation: studies on underlying mechanism of low-level-laser therapy. Med Inf 2015:630361

Wang X, Tian F, Soni SS, Gonzalez-Lima F, Liu H (2016) Interplay between up-regulation of cytochrome-c-oxidase and hemoglobin oxygenation induced by near-infrared laser. Sci Rep 6:30540

Weis S, Rivière R, Deléglise S, Gavartin E, Arcizet O, Schliesser A, Kippenberg TJ (2010) Optomechanically induced transparency. Science 330:1520–1523

Wu Q, Xuan W, Ando T, Xu T, Huang L, Huang YY, Dai T, Dhital S, Sharma SK, Whalen MJ, Hamblin MR (2012) Low-level laser therapy for closed-head traumatic brain injury in mice: effect of different wavelengths. Lasers Surg Med 44:218–226

Yadav A, Gupta A (2017) Noninvasive red and near-infrared wavelength-induced photobiomodulation: promoting impaired cutaneous wound healing. Photodermatol Photoimmunol Photomed 33:4–13

Zang K, Sullivan S, Shanks S (2017) A retrospective study of non-thermal laser therapy for the treatment of toe nail onychomycosis. J Clin Aesthet Dermatol 10:24–30

Zhang RL, Chopp M, Zhang ZG, Jiang Q, Ewing JR (1997) A rat model of focal embolic cerebral ischemia. Brain Res 766:83–92

Zivin JA, Albers GW, Bornstein N, Chippendale T, Dahlof B, Devlin T, Fisher M, Hacke W, Holt W, Ilic S, Kasner S, Lew R, Nash M, Perez J, Rymer M, Schellinger P, Schneider D, Schwab S, Veltkamp R, Walker M, Streeter J, NeuroThera Effectiveness and Safety Trial-2 Investigators (2009) Effectiveness and safety of transcranial laser therapy for acute ischemic stroke. Stroke 40:1359–1364

Adv Exp Med Biol - Clinical and Experimental Biomedicine (2018) 1: 53–57
https://doi.org/10.1007/5584_2018_210
© Springer International Publishing AG, part of Springer Nature 2018
Published online: 9 May 2018

The Timing of Rehabilitation Commencement After Reconstruction of the Anterior Cruciate Ligament

Marek Łyp, Iwona Stanisławska, Bożena Witek, Małgorzata Majerowska, Małgorzata Czarny-Działak, and Ewa Włostowska

Abstract

One of the most common injuries of the knee joint is a rupture of the anterior cruciate ligament (ACL). Most authors believe that early rehabilitation of patients after ACL reconstruction promotes better treatment outcomes. Less is known about the influence of the time that passes from injury to surgical reconstruction. Therefore, the goal of this study was to assess the dependence of treatment outcomes of ACL on injury-to-reconstruction and reconstruction-to-rehabilitation time lags. The study included 30 patients of the mean age 34 ± 7 years with trauma-related rupture of ACL and its surgical reconstruction. The time range from ligament rupture to its reconstruction was 120–180 days and from reconstruction to rehabilitation was 1–120 days. Postsurgical rehabilitation outcomes were assessed with the Lysholm knee scale and the IKDC 2000 subjective knee evaluation form. The scales were applied before and after rehabilitation. We found distinct improvements in all physical symptoms in the damaged knee joint, regardless of the time elapsed from trauma to ACL reconstruction and from ACL reconstruction to rehabilitation. The beneficial outcomes of rehabilitation were significantly inversely associated with the time elapsing from reconstruction to rehabilitation commencement but failed to depend on the time from ACL rupture to reconstruction. We conclude that rehabilitation should start as early as possible after ACL reconstruction to optimize the beneficial outcomes in terms of functional physical recovery, whereas the injury-to-reconstruction delay is less meaningful to this end.

Keywords

Anterior cruciate ligament · Knee joint · Reconstruction · Rehabilitation · Treatment outcome

M. Łyp, I. Stanisławska (✉), M. Majerowska,
and E. Włostowska
Department of Physiotherapy, College of Rehabilitation,
Warsaw, Poland
e-mail: iwona.stanislawska@wsr.edu.pl

B. Witek
Department of Animal Physiology, Institute of Biology,
The Jan Kochanowski University in Kielce, Kielce, Poland

M. Czarny-Działak
Faculty of Medicine and Health Sciences, The Jan
Kochanowski University in Kielce, Kielce, Poland

1   Introduction

The anterior cruciate ligament (ACL) is the most frequently damaged anatomical structure in the knee joint (Anderson et al. 2016; Saka 2014). There are several possible mechanisms of ACL damage. Most often, the injury is caused by crooked or deforming torsional forces acting about the knee joint while the foot is stabilized or by sideward pressure exerted on the loaded limb, with a slight flexion of the knee joint. Young people, actively practicing sport, are most often exposed to ACL injury. Treatment of a ruptured ACL consists of its reconstruction, followed by a comprehensive patient-tailored rehabilitative process, taking into account exercise intensity and pace, extended over several months (Paschos and Howell 2016; Kruse et al. 2012). Rehabilitation programs in individuals with ACL-deficient knees should include proprioceptive and balance exercises, which helps improve outcomes and a return to a full range of knee joint motions (Cooper et al. 2005). A selection of a rehabilitation program depends, to an extent, on coexisting injuries, age, type of activity, and a physical condition of the patient. Studies suggest that post-reconstruction rehabilitation of a patient with ACL injury ought to begin as early as feasible, with the optimum delay of 2–3 days depending on the patient's condition (Grindem et al. 2015; Beynnon et al. 2005). Differences in the effectiveness of rehabilitation have been noticed, depending on the time of its onset after surgical reconstruction (Kochański et al. 2013; Pasierbiński and Jarząbek 2002), but the exact impact of a delayed start of rehabilitative procedures on recovery performance of patients with an ACL injury and their return to full physical activity is unsettled. In this study, we addressed this issue by examining the dependency on the injury-to-ligament reconstruction and reconstruction-to-rehabilitation time delays of the expected beneficial outcomes of rehabilitation after repair of ACL injury. We found that a shortening of the former, but not the latter, associates with outcomes.

2   Methods

This study gained ethical approval from the Human Research Ethics Committee of the College of Rehabilitation in Warsaw, Poland. The study involved 30 patients (20 men and 10 women) of the mean age 34 ± 7 years who suffered unilateral anterior ACL ruptures, followed by arthroscopically assisted reconstruction. The reconstruction consisted of inserting hamstring autographs made with the semitendinosus and gracilis tendons, a double-bundle (STG-DB) technique (Zaffagnini et al. 2006; Fu et al. 2000). Exclusion criteria were previous knee ligament surgery, additional knee injuries, or leg bone fractures. The time range from ligament rupture to surgical reconstruction was from 120 to 180 days (mean 146 ± 96 days) and from reconstruction to rehabilitation was from 1 to 120 days (mean 66 ± 41 days). Rehabilitation was based on cryotherapy and passive and active kinesiotherapy which had all of the patients. In addition, other forms of physiotherapy treatment, such as laser therapy, magnetotherapy, electrotherapy, or patella mobilization, were variably used in some patients. The patients were assessed twice after surgery, before and after rehabilitation using the Tegner Lysholm Knee Scale-Orthopedic Scores and the 2000 International Subjective Knee Evaluation Form (IKDC 2000). The former is a 100-point scale providing information on how the patient's symptoms affect his daily life activities. The scale consists of the following domains: pain (25 points), knee instability (25 points), locking (15 points), swelling (10 points), limping (5 points), stair climbing (10 points), squatting (5 points), and requirement for support (5 points) (Lysholm and Gillquist 1982). The main parts of the latter consisted of patient-reported current health assessment (general ailments) form and sports activities evaluation form (Irrgang et al. 2001). The score for the individual items was summed and then transformed to a scale that ranges from 0 to 100. For both scales, the greater is the score, the fewer symptoms and the better outcome.

Data were displayed as means $\pm$SD and 95% confidence intervals. Differences in the surveyed knee symptoms before and after rehabilitation treatment were assessed with a two-tailed paired *t*-test. Dependence of the treatment outcome on injury-to-reconstruction time and reconstruction-to-rehabilitation time was assessed with Pearson's correlation coefficients. A p-value <0.05 defined statistically significant differences taking place from before to after rehabilitation. The analysis was conducted using a commercial SPSS Statistics software package (IBM Corp.; Armonk, NY).

3   Results and Discussion

In the main, we noticed that the postsurgical rehabilitation had a highly beneficial effect in patients with reconstructed ACL. There were distinct across-the-board improvements in all physical symptoms depicted in both scales used for the assessment of the damaged knee joint, regardless of the time elapsed from trauma to ACL reconstruction and from ACL reconstruction to rehabilitation. Notably, the Lysholm scale shows a 6.5-fold lessening of pain perception, with the tremendous improvement in the ability to climb stairs and disappearing of knee locking symptoms (Table 1). Likewise, the 2000 IKDC scale confirms the improvement in knee joint damage-related declines in muscle strength and endurance and in general physical health, which is most probably related to increased performance of sports activities (Table 2). All these positive changes were substantial as judged from highly significant increases in the scoring of both surveys.

Our present findings confirm those of other recent studies pointing to the importance of physical rehabilitation, in terms of knee function recovery, in patients after ACL reconstruction (Villa et al. 2016; Imoto et al. 2011; Wright et al. 2008; Frańczuk et al. 2004). The main purpose of postoperative rehabilitation is to relieve pain, restore the full function of the knee and the entire limb, and return to a variety of activities as early as feasible (Kochański et al. 2013).

Uncertainty, however, exists about the influence on the effectiveness of rehabilitation of the time scale between the ligament rupture and reconstructive surgery and between the surgery and rehabilitation commencement. In the present study, we attempted to address this issue by seeking the possible association between the two time scales outlined above and the rehabilitation results assessed by the Lysholm and IKDC 2000 scores. We took advantage of the heterogeneity of patients, each having a different circumstance of the ACL injury, health condition, and health care provided thereafter. A dissimilar timeline of treatment procedures enabled the correlation of outcome benefits with the ligament injury-to-reconstruction and reconstruction-to-rehabilitation time lags. We found that all domains of both Lysholm, except knee joint swelling, and IKDC 2000 scales were significantly inversely associated with the time elapsing from reconstructive surgery to rehabilitation commencement, meaning the shorter the delay to

Table 1 Lysholm scale applied in patients before and after rehabilitation of postsurgically reconstructed anterior cruciate ligament (ACL)

Domains	Before rehabilitation	After rehabilitation	$p <$
Pain	2.3 ± 2.9	15.0 ± 8.5	0.001
Knee instability or buckling	6.5 ± 5.1	17.3 ± 4.1	0.001
Knee locking or catching	0.3 ± 0.8	3.3 ± 3.0	0.001
Swelling	1.5 ± 2.0	5.9 ± 1.7	0.001
Limping	0.7 ± 1.3	2.6 ± 1.9	0.001
Stair climbing	0.4 ± 1.1	5.6 ± 3.7	0.001
Squatting	1.1 ± 1.1	3.9 ± 1.9	0.001
Elbow crutches	0.0 ± 0.0	2.7 ± 1.8	0.001
Total score	**12.9 ± 9.6**	**56.1 ± 22.9**	**0.001**

Data are means $\pm$SD

Table 2 The 2000 international subjective knee evaluation form (IKDC 2000) applied in patients before and after rehabilitation of postsurgically reconstructed anterior cruciate ligament

Domains	Before rehabilitation	After rehabilitation	p <
General ailments	11.9 ± 5.5	25.2 ± 10.5	0.001
Sports activities	16.0 ± 6.8	36.9 ± 12.0	0.001
Total score	**27.9 ± 11.0**	**62.2 ± 21.7**	**0.001**

Data are means ±SD

Table 3 Correlation between the surgery-to-rehabilitation time and the results of rehabilitation assessed with the Lysholm and IKDC 2000 scales in patients with injured anterior cruciate ligament (ACL)

Lysholm scale domains	r	p
Pain	−0.418	0.022
Knee instability or buckling	−0.511	0.004
Knee locking or catching	−0.622	<0.001
Swelling	−0.183	0.334
Limping	−0.579	0.001
Stair climbing	−0.837	<0.001
Squatting	−0.694	<0.001
Elbow crutches	−0.406	0.026
Total scale	**0.742**	**<0.001**
IKDC 2000 scale domains		
General ailments	−0.486	0.006
Sports activities	−0.790	<0.001
Total scale	**−0.723**	**<0.001**

r, Pearson's correlation coefficient.

Table 4 Correlation between the injury-to-surgery time and the results of rehabilitation assessed with the Lysholm and IKDC 2000 scales in patients with injured anterior cruciate ligament (ACL)

Lysholm scale domains	r	p
Pain	+0.055	0.773
Knee instability or buckling	+0.050	0.793
Knee locking or catching	+0.259	0.166
Swelling	−0.074	0.697
Limping	−0.101	0.597
Stair climbing	−0.103	0.587
Squatting	−0.055	0.773
Elbow crutches	+0.108	0.571
Total scale	**−0.097**	**0.611**
IKDC 2000 scale domains		
General ailments	−0.097	0.611
Sports activities	−0.004	0.983
Total scale	**−0.050**	**0.792**

r, Pearson's correlation coefficient.

rehabilitation, the better overall physical health outcomes and faster resuming sports activities (Table 3). However, there was no appreciable association between the time elapsing from ACL rupture to reconstructive surgery and the rehabilitation outcomes (Table 4).

In conclusion, we believe we have demonstrated that postoperative rehabilitation should start as early as possible after surgical ACL reconstruction to minimize the effects of injury that caused the ligament rupture. Thus, the present findings lend support to the notion expressed in a recent review of rehabilitation interventions after ACL reconstruction that accelerated rehabilitation may optimize the functional recovery (Grant 2013). On the other hand, we show that a time lag between the injury and undertaking surgical reconstruction is of lesser importance in terms of improved outcome of subsequent rehabilitation. Nonetheless, aggressive rehabilitation does not always bring the intended

results. ACL is a sensitive ligament, and too early loading of it can lead to a re-injury (Stańczak et al. 2014). Individually targeted rehabilitation process in different patients, taking into account specific patient-oriented rehabilitation factors, may play a key role in maximizing the expected postsurgical outcomes.

Conflicts of Interest The authors declare no conflicts of interest in relation to this article.

References

Anderson MJ, Browning WM 3rd, Urband CE, Kluczynski MA, Bisson LJ (2016) A systematic summary of systematic reviews on the topic of the anterior cruciate ligament. Orthop J Sports Med 4 (3):2325967116634074

Beynnon BD, Uh BS, Johnson RJ, Abate JA, Nichols CE, Fleming BC, Poole AR, Ross H (2005) Rehabilitation after anterior cruciate ligament reconstruction: a pro- spective, randomized, double-blind comparison of programs administered over 2 different time intervals. Am J Sports Med 33(3):347–359

Cooper RL, Taylor NF, Feller JA (2005) A systematic review of the effect of proprioceptive and balance exercises on people with an injured or reconstructed anterior cruciate ligament. Res Sports Med 13 (2):163–178

Frańczuk B, Fibiger W, Kukiełka R, Jasiak-Tyrkalska B, Trąbka R (2004) Early rehabilitation after arthroscopic reconstruction of the anterior cruciate ligament. Ortop Traumatol Rehab 6(4):416–422

Fu FH, Bennett CH, Ma CB, Menetrey J, Menetrey J, Lattermann C (2000) Current trends in anterior cruciate ligament reconstruction: Part II. Operative procedures and clinical correlations. Am J Sports Med January 28:124–130

Grant JA (2013) Updating recommendations for rehabili- tation after ACL reconstruction: a review. Clin J Sport Med 23(6):501–502

Grindem H, Granan LP, Risberg MA, Engebretsen L, Snyder-Mackler L, Eitzen I (2015) How does a com- bined preoperative and postoperative rehabilitation programme influence the outcome of ACL reconstruc- tion 2 years after surgery? A comparison between patients in the Delaware-Oslo ACL Cohort and the Norwegian National Knee Ligament Registry. Br J Sports Med 49(06):385–389

Imoto AM, Peccin S, Almeida GJ, Saconato H, Atallah ÁN (2011) Effectiveness of electrical stimulation on rehabilitation after ligament and meniscal injuries: a systematic review. Sao Paulo Med J 129(6):414–423

Irrgang JJ, Anderson AF, Boland AL Harner CD, Kurosaka M, Neyret P, Richmond JC, Shelborne KD (2001) Development and validation of the international knee documentation committee subjective knee form. Am J Sports Med 29:600–613

Kochański B, Łabejszo A, Kałużny K, Mostowska K, Wołowiec Ł, Trela E, Hagner W, Zukow W (2013)

Knee injury – a diagnostic procedure. J Health Sci 3 (5):439–456

Kruse LM, Gray B, Wright RW (2012) Rehabilitation after anterior cruciate ligament reconstruction: a systematic review. J Bone Joint Surq Am 94(19):1737–1748

Lysholm J, Gillquist J (1982) Evaluation of knee ligament surgery results with special emphasis on use of a scor- ing scale. Am J Sports Med 10:150–154

Paschos NK, Howell SM (2016) Anterior cruciate liga- ment reconstruction: principles of treatment. Efort Open Rev 1(11):398–408

Pasierbiński A, Jarząbek A (2002) Rehabilitation after anterior cruciate ligament reconstruction. Acta Clinica 2(1):86–100

Saka T (2014) Principles of postoperative anterior cruciate ligament rehabilitation. World J Orthop 5(4):450–459

Stańczak K, Domżalski M, Synder M, Sibiński M (2014) Return to motor activity after anterior cruciate ligament reconstruction - pilot study. Ortop Traumatol Rehabil 5 (6):477–486

Wright RW, Preston E, Fleming BC, Amendola A, Andrish JT, Bergfeld JA, Dunn WR, Kaeding C, Kuhn JE, Marx RG, McCarty EC, Parker RC, Spindler KP, Wolcott M, Wolf BR, Williams GN (2008) A systematic review of anterior cruciate ligament recon- struction rehabilitation: part I: continuous passive motion, early weight bearing, postoperative bracing, and home-based rehabilitation. J Knee Surg 21 (3):217–224

Villa FD, Ricci M, Perdisa F, Filardo G, Gamberini J, Caminati D, Villa SD (2016) Anterior cruciate liga- ment reconstruction and rehabilitation: predictors of functional outcome. Joints 3(4):179–185

Zaffagnini S, Marcacci M, Lo Presti M, Giordano G, Iacono F, Neri MP (2006) Prospective and randomized evaluation of ACL reconstruction with three techniques: a clinical and radiographic evaluation at 5 years follow-up. Knee Surg Sports Traumatol Arthrosc 14:1060–1069

Adv Exp Med Biol - Clinical and Experimental Biomedicine (2018) 1: 59–64
https://doi.org/10.1007/5584_2018_209
© Springer International Publishing AG, part of Springer Nature 2018
Published online: 9 May 2018

Novel Model of Somatosensory Nerve Transfer in the Rat

Adriana M. Paskal, Wiktor Paskal, Kacper Pelka,
Martyna Podobinska, Jaroslaw Andrychowski,
and Pawel K. Wlodarski

Abstract

Nerve transfer (neurotization) is a reconstructive procedure in which the distal denervated nerve is joined with a proximal healthy nerve of a less significant function. Neurotization models described to date are limited to avulsed roots or pure motor nerve transfers, neglecting the clinically significant mixed nerve transfer. Our aim was to determine whether femoral-to-sciatic nerve transfer could be a feasible model of mixed nerve transfer. Three Sprague Dawley rats were subjected to unilateral femoral-to-sciatic nerve transfer. After 50 days, functional recovery was evaluated with a prick test. At the same time, axonal tracers were injected into each sciatic nerve distally to the lesion site, to determine nerve fibers' regeneration. In the prick test, the rats retracted their hind limbs after stimulation, although the reaction was moderately weaker on the operated side. Seven days after injection of axonal tracers, dyes were visualized by confocal microscopy in the spinal cord. Innervation of the recipient nerve originated from higher segments of the spinal cord than that on the untreated side. The results imply that the femoral nerve axons, ingrown into the damaged sciatic nerve, reinnervate distal targets with a functional outcome.

Keywords

Axonal tracers · Nerve fibers · Nerve transfer · Neuroregeneration · Neurotization · Sciatic nerve · Spinal cord injury

Electronic supplementary material: Supplementary material is available in the online version of this chapter at https://doi.org/10.1007/5584_2018_209.

A. M. Paskal and W. Paskal
Department of Experimental and Clinical Physiology, Laboratory of Center for Preclinical Research, Warsaw Medical University, Warsaw, Poland

Department of Histology and Embryology, Laboratory of Center for Preclinical Research, Warsaw Medical University, Warsaw, Poland

K. Pelka and P. K. Wlodarski (✉)
Department of Histology and Embryology, Laboratory of Center for Preclinical Research, Warsaw Medical University, Warsaw, Poland
e-mail: pawel.wlodarski@wum.edu.pl

M. Podobinska
Department of Experimental and Clinical Physiology, Laboratory of Center for Preclinical Research, Warsaw Medical University, Warsaw, Poland

J. Andrychowski
Department of Neurology and Neurosurgery, Faculty of Medical Sciences, University of Warmia and Mazury in Olsztyn, Olsztyn, Poland

1   Introduction

Spinal cord injury (SCI) is an irreversible chronic condition vastly impairing patient's quality of life and independence. SCI at the cervical level resulting in tetraplegia is especially devastating, since essential hand functions are lost or greatly impaired. In some cases, motor functions of the upper limb can be partially restored by reconstructive procedures, i.e., neurotizations (nerve transfers). A neurotization procedure aims to revive the function of a distal denervated nerve element by connecting it with a proximal healthy donor nerve of a less significant function that can be sacrificed. As a result, the donor nerve axons grow into the recipient nerve pathway and innervate its distal targets such as muscles and skin. A major requirement is that the expected outcome of the procedure ought to result in a significant gain of function in the recipient nerve and a minimal loss of function resulting from the loss of function of a donor nerve. Specific qualification criteria for neurotization surgery have been described by Senjaya and Midha (2013). With respect to the upper limb, the following functions can be regained: elbow flexion, thumb opposition and extension, and finger flexion and extension. Apart from motor function recovery, neurotization technique also plays a role in treating recurrent pressure ulcers by restoring sensory innervation of the altered region. In this case, skin and subcutaneous tissue are distal targets for the donor nerve (Viterbo and Ripari 2008; Louie et al. 1987). Nerve transfers also are used to restore bladder control (Gomez-Amaya et al. 2015; Xiao et al. 2003) and for diaphragm reanimation (Krieger and Krieger 2000).

Other than the functional assessment of motor and sensory function and electrophysiological examinations, the methods of a molecular analysis of neurotization in humans are limited, and the currently available animal models have serious limitations. There are few examples of neurotizations performed in animals in the literature such as a rabbit model of neurotization within the brachial plexus (Cao et al. 2003), a cat model

of accessory-to-ulnar nerve neurotization (Isla et al. 2006), a sheep model of inter-costo-lumbar neurotization (Vialle et al. 2010), a rat model of long thoracic-to-thoracodorsal nerve neurotization (motor nerve transfer) (Spyropoulou et al. 2007), a rat model of sensory-to-motor nerve neurotization (Noordin et al. 2008), and a rat model of ipsilateral cervical nerve root transfer (Song et al. 2010). The reported models of neurotizations, due to a need for a complex management, are performed in higher vertebrates, and the results are hardly reproducible. The available data concerning neurotization in rat models by far focus on root avulsion or selective motor nerve transfers. To date, there is no research in rats concerning mixed nerve transfers (motor-to-sensory) in which the effect of neurotization would be verified by neuronal retrograde tracing. Mixed nerve transfers are liable to be more clinically relevant, as the two components of innervation are crucial for patient recovery. Only has a single study employed mixed nerve transfer in a rat model of upper limb paralysis treatment, but the mechanism of reinnervation is not fully explained (Rodriguez et al. 2011). There is a need to develop a feasible, reproducible animal model of neurotization, which would allow the exploration of molecular aspects of neuroregeneration of a peripheral nerve into a foreign nerve pathway, neural plasticity, and regulatory factors.

In the present study, we addressed this issue by designing a model of femoral-to-sciatic nerve neurotization in a rat model, with the possible aim to restore both motor and sensory functions of a lower limb. A model of mixed spinal nerve, composed of motor and sensory fascicles, transfer would mimic the most frequent type of nerves in the human body. To verify the concept of neurotizing a nerve with major innervation with a nerve with minor innervation, we chose the sciatic and femoral nerves, respectively. The rat sciatic nerve (spinal origin L4–L6) innervates the majority of the hind limb muscles, while the femoral nerve (spinal origin L2–L4) innervates the anterior thigh muscle only (*quadriceps femoris* muscle). To visualize the level of origin of spinal neurons following neurotization, we labeled the nerves with retrograde fluorescent axonal tracers.

2 Methods

2.1 Animals

The experiments were approved by the First Local Ethics Committee in Warsaw (permit no. 099/2016). Care and handling of all animals were carried out in accordance with the Act 1986 and associated guidelines of UK's Animals Scientific Procedures and the 2010/63/EU Directive for animal experiments and complied with the ARRIVE guidelines. The rats were purchased from the Central Laboratory of Experimental Animals of the Medical University of Warsaw (license no. 0037). All surgical procedures were performed using aseptic techniques. Inbred male Sprague Dawley rats, weighing 270–295 g, aged 8–10 weeks (n = 3), were acclimatized to 12 h light/dark cycle at 19 °C, with water and normal food ad libitum. Following surgery, the rats were housed in separate cages to avoid biting of the denervated leg by cohabitants.

2.2 Experimental Design

On Day 0, we performed surgical femoral-to-sciatic nerve transfer through the left inguinal approach. The animals were observed for 50 days after surgery. On day 50, a prick test of each hind limb was taken. Next, animals were anesthetized, and the retrograde fluorescent axonal tracer was injected into each sciatic nerve at the thigh level. Seven days later, animals were euthanized and perfused with a fixative solution. The spinal cord was dissected from the vertebral canal for a microscopic examination. Figure 1 summarizes the study design.

2.3 Femoral-To-Sciatic Nerve Transfer Procedure

One hour before induction of anesthesia, the rats received an injection of the analgesic buprenorphine chloride (3 μg/100 g; i.p.) (Bupaq Multidose, Richter Pharma AG; Wels, Austria). The animals were anesthetized with ketamine (100 mg/kg; i.p.) and xylazine (10 mg/kg; i.p.) (Vetoquinol Biowet; Pulawy, Poland). A 3-cm incision was aseptically made along the left inguinal sulcus. The exposed neurovascular bundle contained the femoral nerve, a donor nerve for neurotization. The nerve was separated from the bundle and followed proximally. The inguinal ligament was cut, and the longissimus lumborum muscle was divided to expose the sciatic nerve, a recipient nerve for neurotization, which was separated from the surrounding structures. To provide tension-free linkage, the nerves were cut. The femoral nerve was cut as far distally and the sciatic nerve as far proximally as possible. The two nerves were joined with four epineurial 10-0 nylon microsutures. The wound was irrigated with saline, and the neural anastomosis was hidden under the muscles which were sutured. The skin was closed with 4-0 Vicryl sutures (Yavo, Belchatów, Poland). All of the surgical procedures, after the exposure of the neurovascular bundle, were performed with the aid of an operative microscope at 30× magnification.

The age of rats used in the experiment was less than 10 weeks. Older rats cannot be used as the longissimus lumborum muscle volume may be too large. Thus, distance between the sciatic and femoral nerves would be too long, which might hinder a tension-free nerve transfer.

2.4 Postoperative Management

The animals were observed for 50 days after surgery. On postoperative day 50, a prick test was performed, with a needle prick of the plantar surface of either foot. The nonoperated limb was pricked first to observe a reaction, followed by the limb operated on.

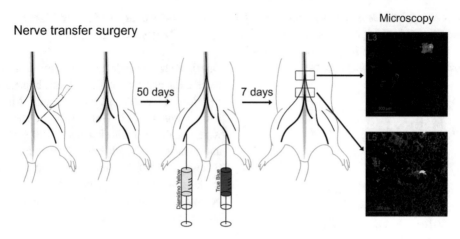

Fig. 1 Scheme of a rat model of femoral-to-sciatic nerve transfer

Afterward, animals were anesthetized with ketamine (100 mg/kg) and xylazine (10 mg/kg; i.p.), and 1-cm-long skin incisions were made on the lateral side of both thighs 1 cm below and parallel to the femur. The thigh muscles were split to expose the sciatic nerve. Using a Hamilton syringe, we injected retrograde fluorescent axonal tracers 3 mm above each sciatic nerve's trifurcation. The "diamidino yellow" tracer was injected in the nonoperated limb and "true blue" tracer into the operated limb, 1 μL of 2% solution each. Seven days after the injection, animals were sacrificed with sodium pentobarbital (1 mL/kg) (Biowet, Pulawy, Poland) and perfused intracardially with heparin (1,000 IU/kg in 100 mL physiological saline), 4% paraformaldehyde (500 mL), and 200 mL 10% sucrose in phosphate buffer (pH 7.40). The spinal cord was harvested, divided into anatomical segments, and post-fixed in 4% paraformaldehyde for 3 h at 4 °C and then immersed in 15% sucrose in phosphate buffer at 4 °C overnight. The tissue was then embedded in the optimal cutting temperature compound, cut into 30 μm slices in a cryostat (Leica, Wetzlar, Germany), and visualized with a 710 Zeiss NLO confocal microscopy equipped with a multiphoton unit (Chameleon Ultra–Coherent, Santa Clara, CA).

"Diamidino yellow" labels neuronal nuclei (excitation, 365 nm; emission, > 500 nm), and "true blue" labels the neuronal cytoplasm, nucleoli, proximal dendrites, and axons (excitation, 365 nm; emission, 405 nm). The images were assessed by a researcher blinded to the study purpose.

3   Results

3.1   Motor and Sensory Function Recovery

Comparison of a sensorimotor response to a prick test in the nonoperated, control hind limb with that in the other hind limb after a surgical femoral-to-sciatic nerve transfer is shown in Video 1. Function, assessed from the limb retraction on prick, recovered after the nerve transfer, albeit it was weaker compared with that in the control condition.

3.2   Microscopic Analysis

Confocal microscopy showed the presence of retrograde axonal fluorescent tracers within different

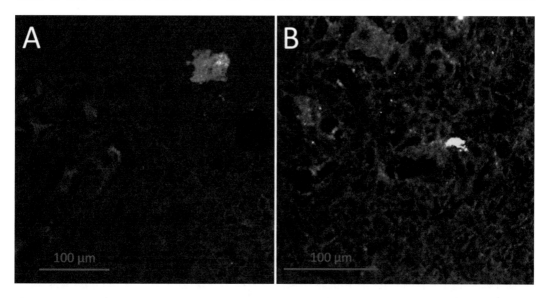

Fig. 2 Confocal microscopy images: (**a**) "true blue" tracer in L3 spinal cord section; (**b**) "diamidino yellow" in L5 spinal cord section

spinal cord segments. The "true blue" tracer was visualized in the cytoplasm of neural cells in the ventral horn at L3 and L4 levels, while "diamidino yellow" was in the nuclei of neural cells in the ventral horn at L5 and L6 levels (Fig. 2a, b, respectively).

4 Discussion

In the era when the development of medical science is at the highest pace, reconstructive surgery techniques are no longer part of science fiction but are being used in animal models and will eventually be used on humans. One such innovative solution concerning paralyzed patients is the nerve transfer technique. Instead of replacing denervated muscle functions by tendon transfers, an attempt is made to innervate the desired targets (significant muscles to decrease patient dependence or skin areas endangered by the development of pressure ulcers) (Senjaya and Midha 2013; Viterbo and Ripari 2008). Nerve transfers appear to be the future of restorative medicine, and every attempt to understand better the

neuroregeneration process within neurotization has to be made. The molecular aspects of neurotizations, often hard or impossible to examine in humans, should be explored with the aid of animal models in order to enhance this technique. We encourage researchers to use our rat model of femoral-to-sciatic nerve transfer due to its confirmed feasibility and simplicity. A better understanding of the basic mechanisms of neuroregeneration within neurotization may increase the number of these procedures in humans. Nonetheless, the decision to qualify a patient for this technique should be taken with caution, taking into careful consideration the individual patient's anatomical situation and the potential consequences of the loss of function of the chosen donor nerve (Senjaya and Midha 2013).

5 Conclusions

Restoration of sensorimotor function in the operated limb demonstrates that the femoral nerve can substitute for the sciatic nerve to innervate its distal targets with a satisfactory outcome. The tracers found in the spinal cord in the present study corresponded with the clinical sensorimotor function as assessed by the prick test. Intraneural injections of the retrograde fluorescent axonal tracers into the sciatic nerve resulted in a retrograde axonal transport of tracers to perikaryons; "diamidino yellow" was transported into nuclei and "true blue" into the cytoplasm. A "diamidino yellow" solution injected into the nonoperated sciatic nerve pointed to the L5–L6 spinal origin of the sciatic nerve motoneurons, with the denser number of the dye positive cells at the L5 level. On the other hand, "true blue" solution injected into the operated sciatic nerve confirmed the presence of the primary femoral nerve axons in the sciatic nerve pathway. We observed the highest number of "true blue" positive cells at the L3 level, corresponding to the femoral nerve origin, on the operated side. A rat model of femoral-to-sciatic nerve transfer appeared a viable way to explore molecular aspects of neuroregeneration in the neurotization process. Although our study was limited to three rats in order to demonstrate the novel technique, wc bclicvc we have shown the presence of reinnervation and neural plasticity emanating from a foreign donor nerve.

Acknowledgments Supported by the Medical University of Warsaw, Second Faculty of Medicine, Grant No. 2 W8/NM2/16, and the statutory funds from the First Faculty of Medicine, Medical University of Warsaw. The funder had no involvement in the study design nor in data collection, analysis, or interpretation.

Conflicts of Interest The authors declare no conflicts of interest in relation to this article.

References

Cao X, Li J, Cao Y, Cai J (2003) C3, 4 transfer for neurotization of C5, 6 nerve roots in brachial plexus injury in a rabbit model. J Reconstr Microsurg 19 (4):265–270

Gomez-Amaya SM, Barbe MF, de Groat WC, Brown JM, Tuite GF, Corcos J, Fecho SB, Braverman AS, Ruggieri MR Sr (2015) Neural reconstruction methods of restoring bladder function. Nat Rev Urol 12 (2):100–118

Isla A, Martinez JR, Perez Lopez C, Perez Conde C, Morales C, Avendano C (2006) Anatomical and functional connectivity of the transected ulnar nerve after accessory nerve neurotization in the cat. J Neurosurg Sci 50(2):33–40

Krieger LM, Krieger AJ (2000) The intercostal to phrenic nerve transfer: an effective means of reanimating the diaphragm in patients with high cervical spine injury. Plast Reconstr Surg 105(4):1255–1261

Louie G, Mackinnon SE, Dellon AL, Patterson GA, Hunter DA (1987) Medial antebrachial cutaneous--lateral femoral cutaneous neurotization in restoration of sensation to pressure-bearing areas in a paraplegic: a four-year follow-up. Ann Plast Surg 19(6):572–576

Noordin S, Ahmed M, Rehman R, Ahmad T, Hashmi P (2008) Neuronal regeneration in denervated muscle following sensory and muscular neurotization. Acta Orthop 79(1):126–133

Rodriguez A, Chuang DC, Chen KT, Chen RF, Lyu RK, Ko YS (2011) Comparative study of single-, double-, and triple-nerve transfer to a common target: experimental study of rat brachial plexus. Plast Reconstr Surg 127(3):1155–1162

Senjaya F, Midha R (2013) Nerve transfer strategies for spinal cord injury. World Neurosurg 80(6):e319–e326

Song J, Chen L, Gu Y (2010) Effect of ipsilateral C7 nerve root transfer on restoration of rat upper trunk muscle and nerve function after brachial plexus root avulsion. Orthopedics 33(12):886

Spyropoulou GA, Lykoudis EG, Batistatou A, Papalois AE, Tagaris G, Pikoulis E, Bastounis E, Papadopoulos O (2007) New pure motor nerve experimental model for the comparative study between end-to-end and end-to-side neurorrhaphy in free muscle flap neurotization. J Reconstr Microsurg 23(7):391–398

Vialle R, Lacroix C, Harding I, Loureiro MC, Tadie M (2010) Motor and sensitive axonal regrowth after multiple intercosto-lumbar neurotizations in a sheep model. Spinal Cord 48(5):367–374

Viterbo F, Ripari WT (2008) Nerve grafts prevent paraplegic pressure ulcers. J Reconstr Microsurg 24 (4):251–253

Xiao CG, Du MX, Dai C, Li B, Nitti VW, de Groat WC (2003) An artificial somatic-central nervous system-autonomic reflex pathway for controllable micturition after spinal cord injury: preliminary results in 15 patients. J Urol 170(4 Pt 1):1237–1241

Adv Exp Med Biol - Clinical and Experimental Biomedicine (2018) 1: 65–71
https://doi.org/10.1007/5584_2018_190
© Springer International Publishing AG, part of Springer Nature 2018
Published online: 29 March 2018

Next-Generation Sequencing of Hepatitis C Virus (HCV) Mixed-Genotype Infections in Anti-HCV-Negative Blood Donors

Maciej Janiak, Kamila Caraballo Cortés, Karol Perlejewski, Dorota Kubicka-Russel, Piotr Grabarczyk, Urszula Demkow, and Marek Radkowski

Abstract

The infection with more than one hepatitis C virus (HCV) genotype especially in subjects with a high risk of multiple HCV exposures has been demonstrated. The role of HCV mixed-genotype infection in viral persistence and treatment effect is not fully understood. The prevalence of such infection varies greatly depending on the technique used for genotype determination and studied population. Next-generation sequencing (NGS) which is suitable for extensive analysis of complex viral populations is a method of choice for studying mixed infections. The aim of the present study was to determine the prevalence of mixed-genotype HCV infections in the Polish sero-negative, HCV-RNA-positive blood donors (n = 76). Two-step PCR was used for amplification of 5′-UTR of HCV. Using pyrosequencing altogether, 381,063 reads were obtained. The raw reads were trimmed and subjected to similarity analysis against the entire unfiltered NCBI nt database. Results obtained from NGS were compared with the standard genotyping. One (1.3%) mixed-genotype [3a, 2989 reads (94.8%); 1b, 164 reads (5.2%)] infection was found in a sample diagnosed as genotype 3a only by routine testing. Two samples were identified with different genotypes, compared to routine testing. In conclusion, NGS is a sensitive method for HCV genotyping. The prevalence of mixed-genotype HCV infections in blood donors is low.

Keywords

Blood donors · HCV genotyping · HCV mixed-genotype infections · Hepatitis C virus · Next-generation sequencing · Pyrosequencing

1 Introduction

Hepatitis C virus (HCV) infection is recognized worldwide a major health issue (Huang et al. 2015). However, distribution of HCV displays a

M. Janiak, K. Caraballo Cortés (✉), K. Perlejewski, and M. Radkowski
Department of Immunopathology of Infectious and Parasitic Diseases, Medical University of Warsaw, Warsaw, Poland
e-mail: kcaraballo@wum.edu.pl

D. Kubicka-Russel and P. Grabarczyk
Department of Virology, Institute of Hematology and Transfusion Medicine, Warsaw, Poland

U. Demkow
Department of Laboratory Medicine and Clinical Immunology of Developmental Age, Medical University of Warsaw, Warsaw, Poland

considerable variation across regions and countries (Sakem et al. 2016). According to recent estimations, there are 115 M persons positive for hepatitis C antibodies (anti-HCV) and 80 M HCV viremia individuals in the world (Madalinski et al. 2015). Seven genotypes of HCV are distinguished by phylogenetic methods (Bagaglio et al. 2015) and more than 50 subtypes with tendency to segregate geographically (Huang et al. 2015). Genotype 1 is the most prevalent worldwide and accounts for 70% of infections in the USA, while genotypes 2 and 3 are responsible for the majority of the remaining 30%. Genotype 2 occurs widely in Central Africa, while genotype 4 dominates in the Middle East, genotype 5 in South Africa, and genotype 6 in South China and Southeast Asia.

The predominance of genotype 1 in Poland may be explained by the fact that before 1992 a principal route of HCV transmission were blood products (Gowin et al. 2016). During the period of 2002–2012, a number of serological surveys were conducted in Poland to establish the prevalence of HCV infections (Sakem et al. 2016). The current transmission of HCV through medical procedures is considered low (Gowin et al. 2016). The estimated anti-HCV prevalence is 0.86%, while HCV-RNA positivity indicating "active" infection is detectable in 0.6% of the population (Sakem et al. 2016). The distribution of genotypes in Poland displays the predominance of genotype 1 (79.4%), followed by genotype 3 (13.8%) and genotype 4 (4.9%) (Panasiuk et al. 2013).

Co-infection of the same patient with more than one HCV genotype has been described (Hnatyszyn 2005; Giannini et al. 1999). In the general population, mixed-genotype infection was reported to be present in 2–10% of all HCV-positive patients (Butt et al. 2011; Zarkesh-Esfahani et al. 2010; Schroter et al. 2003). However, inconsistent data regarding a real prevalence of this phenomenon are observed, suggesting a bias due likely to demographic and methodologic factors (Giannini et al. 1999). The prevalence of HCV mixed infections is variable when the infection is tested by different assays in the same group of patients (Parodi et al. 2008).

In the high-risk populations, such as the hemodialyzed, hemophiliacs, and the patients with von Willebrand's disease, the frequency of mixed infections is about 13% (Bagaglio et al. 2015; Hairul Aini et al. 2012; Qian et al. 2000), while in intravenous drug users and men who have sex with men, the incidence is from 25% to 39% (Gowin et al. 2016). Nonetheless, in some groups of parenterally transmitted infection, like beta-thalassemia patients and HIV-infected intravenous drug users, the frequency of mixed-genotype infection is even higher, up to 50.8% (Bagaglio et al. 2015). It is suggested that mixed-genotype HCV infections may be much more common in the general population than anticipated and may represent a factor associated with clinical course of disease and response to antiviral therapy (Hnatyszyn 2005).

The assessment of a real prevalence of HCV mixed-genotype infections is hampered by the use of low-sensitivity techniques, suitable for identifying a dominant HCV genotype only. Conventional sequencing methods and genotyping based on PCR amplification of conserved regions, with universal primers such as restriction fragment length polymorphism (RFLP), have limited sensitivity and may not detect minor variants (Caraballo Cortes et al. 2016; Giannini et al. 1999). In an extensive study, Hu et al. (2000) have compared the currently available genotyping assays, such as type-specific polymerase chain reaction (T-S PCR), RFLP analysis, and the line probe assay (LiPA), using the samples collected from randomly selected HCV-positive blood donors, thalassemia patients, drug users, and persons with chronic liver disease. The results show significant inter-method differences in the detection rates of mixed genotypes, e.g., 35.7%, 16.7%, and 14.3% for PS-PCR, LiPA, and RFLP analysis, respectively.

Higher sensitivity of HCV genotype detection may be achieved by the next-generation sequencing (NGS). Due to the possibility of generating millions of reads in a single run, detection of minor viral variants is possible (Caraballo Cortes et al. 2016). The aim of this study was to determine the prevalence of HCV mixed-genotype infections in the Polish blood donors

(HCV-RNA-positive, anti-HCV-negative) using pyrosequencing and to compare the results with those obtained by a standard genotyping method.

2 Methods

2.1 Patients and Samples

The study involved a panel of 99 serum samples collected between 1999 and 2015 from HCV-RNA-positive, anti-HCV-negative blood donors infected with genotypes 1 and 3: 1a, 5 (5.1%); 1b, 42 (42.4%); and 3a, 52 (52.5%). The genotype was determined with use of a Versant HCV Genotype assay or Versant HCV Genotype 2.0 assay (LiPA) (Siemens AG, Munich, Germany) in the Institute of Hematology and Transfusion Medicine in Warsaw, Poland, as part of routine procedures. All subjects had been screened during 6 months preceding the sampling for HCV-RNA and anti-HCV in the Regional Blood Transfusion Centers with commercially available tests and were found negative. No human immunodeficiency virus (HIV) or hepatitis B virus (HBV) co-infected subjects were included in the study.

2.2 HCV-RNA Extraction and 5′UTR Amplification

Total RNA was extracted from 250 µl of serum by a modified guanidinium thiocyanate phenol/chloroform method using Trizol (Life Technologies, Carlsbad, CA). 5.65 µL RNA was subjected to reverse transcription at 42 °C for 60 min using AccuScript High Fidelity Reverse Transcriptase (Agilent Technologies, Santa Clara, CA) and Random Hexamers (Invitrogen, Carlsbad, CA).

A two-step PCR was used for amplification of 5′-UTR of HCV. Primers used in the first step were as follows: 5′-TGRTGCACGGTCTACGA-GACCTC-3′ (nt 342–320) and 5′-RAYCA-CTCCCCTGTGAGGAAC-3′ (nt 33–55). PCR was carried out using 2.5 U Fast Start High Fidelity Enzyme Blend (Roche Diagnostics, Indianapolis, IN), 1x buffer, 1.8 mM MgCl$_2$,

0.2 mM dNTP, and 40 pmol of each primer in a total volume of 50 µL. Two microliters of cDNA were added to the PCR mixture. Thermal profile for the amplification was as follows: initial denaturation at 94 °C for 5 min, 50 cycles with denaturation at 94 °C for 1 min, and annealing at 58 °C for 1 min. The reaction was terminated with a final elongation at 72 °C for 7 min.

The second-step PCR was performed with fusion primers containing multiplex identifiers (MID) specific for each sample. The sequence-specific primers were as follows: forward primer 5′-ACTGTCTTCACGCAGAAAGCGTC-3′ and reverse primer 5′-CAAGCACCCTATCAGGC-AGTACC-3′. Amplicon products were extracted from agarose gel by QIAquick Gel Extraction Kit (Qiagen, Hilden, Germany), purified by A gencourt AMPure XP purification system (Beckman Coulter, Beverly, MA), and measured fluorometrically using Quant-iT™ High-Sensitivity dsDNA Assay Kit (Life Technologies, Carlsbad, CA) on Qubit 3.0 Fluorometer (Life Technologies, Carlsbad, CA).

2.3 Pyrosequencing

The amount of DNA equivalent to 3×10^7 amplicons was subjected to emulsion PCR using GS Junior Titanium emPCR Lib-A Kit (454/Roche, Branford, CT). Pyrosequencing was performed following the amplicon processing protocol according to the manufacturer's instruction using a GS Junior System (454/Roche, Branford, CT).

2.4 Bioinformatic Analysis

Raw reads were trimmed using cutadapt-1.2.1 (Martin 2011) resulting in approximately 250 bp sequences. Trimmed bases with quality below Q20 (Phred quality score) were removed using fastx_artifacts_filter. Identical reads were grouped, counted, and ordered from the most to least abundant. Unique sequences were aligned using Clustal X v2.0 (Larkin et al. 2007). Then, reads were subjected to similarity search using blastn (Altschul et al. 1990) against the entire unfiltered NCBI nt database with e-value cutoff of $1e^{-5}$.

3   Results

Using NGS, 381,063 5′UTR reads were obtained from 76 samples out of the initial 99 sample collection. The remaining samples were discarded due to the lack of amplification or sequencing results. The characteristics of the study group are presented in Table 1. The number of reads ranged from 413 to 14,435 *per* sample, and the average number of reads *per* sample was 5337,

Table 1 Characteristics and next-generation sequencing (NGS) genotyping results of blood donors diagnosed as HCV-RNA-positive and anti-HCV-negative

Sex, n (%)	
Male	65 (85.5)
Female	11 (14.5)
Age (year)	
Mean	28.6
Median	26.5
Range	18–55
Viral load (IU/mL)	
Mean	4,045,728
Median	560,500
Range	184–6,900,000
Alanine aminotransferase (ALT) (IU/L)	
Mean	61.9
Median	46.5
Range	9–500
Genotype distribution[a] n (%)	
1a	4 (5.2)
1b	36 (47.4)
3a	36 (47.4)
Genotype distribution[b] n (%)	
1a	3 (3.9)
1b	37 (48.7)
3a	34 (44.8)
4a	1 (1.3)
3a/1b	1 (1.3)

[a]According to Versant HCV Genotype assay or Versant HCV Genotype 2.0 assay (LiPA) (Siemens AG, Munich, Germany)
[b]According to sequencing by GS Junior System (454/Roche, Branford, CT)

with a median of 4644. All samples were routinely evaluated as harboring single HCV genotype.

In one sample (1.3%), diagnosed as genotype 3a only, NGS detected two genotypes: 3a and 1b. The number of 3a genotype reads was 2989 representing 94.8% of the entire sequence population, whereas genotype 1b was represented by 164 reads which constituted 5.2% of frequency (Fig. 1). In one sample routinely determined as genotype 1a, NGS showed 1b. In another sample routinely diagnosed as genotype 3a, NGS revealed the presence of genotype 4a.

4   Discussion

The mixed-genotype HCV infections result from co-infection or superinfection with different viral genotypes or subtypes. This phenomenon may have clinical importance affecting the natural course of infection, viral load, dynamics of transmission, and treatment response. Some authors have observed that this form of infection is often accompanied by increased HCV-RNA and aminotransferase activity. Moreover, infection with more than one genotype may result in development of more virulent recombinant strains (Caraballo Cortes et al. 2016). Superinfection with an another genotype can be responsible for exacerbation of chronic HCV infection (Kao et al. 1994), and patients with mixed-genotype infection seem to experience more severe liver damage (Widell et al. 1995). On the other hand, some authors have not observed significant differences of HCV viral loads between persons infected with a single HCV genotype and those infected with multiple HCV genotypes (Schijman et al. 2004).

The identification of HCV genotype can be performed using amplification of specific regions and type-specific determination of sequence differences by restriction fragment length polymorphism (RFLP), sequence analysis, or line probe reverse hybridization (Hu et al. 2000; Buoro et al. 1999). The routinely used methods are suitable only for detecting a dominant genotype strain, e.g., RFLP analysis is able to detect genotype constituting $\geq 41.6\%$ in a mixed-

```
3a(1101)  CAAGCACCCTATCAGGCAGTACCACAAGGCCTTTCGCGACCCAACACTACTCGGCTAGTGATCTCGCGGGGCACGCCCAAATTTCTGGGTATTGAGCGGGTTGTTCCAAGAAAGGACCCGGTCAC
3a(509)   .............................................................................................................................
3a(231)   .............................................................................................................................
3a(151)   .............................................................................................................................
3a(54)    .............................................................................................................................
1b(39)    ..........................................CAG...............C..CA..C...............A.....................T
1b(15)    ...........................................................C..CA..C...............A.....................T
1b(10)    ..........................................CAG...............C..CA..C...............A.....................T
1b(8)     ..........................................CAG...............C..CA..C...............A.....................T
1b(6)     ..........................................CAG...............C..CA..C...............A.....................T

3a(1101)  CCCAGCGATTCCGGTGTACTCACCGGTTCCGCAGACCACTATGGCTCTCCCGGGAGGGGGGGTCCTGGAGGCTGCACGACACTCGTACTAACGCCATGGCTAGACGCTTTCTGCGTGAAGACAGT
3a(509)   ...................................................................................G...........................
3a(231)   .......................................A...................A...................
3a(151)   .......................................A...................A...................
3a(54)    .......................................A...................-...................
1b(39)    ..TG..A................................A...................A...................
1b(15)    ..TG..A................................A...................A...................
1b(10)    .............................................................................
1b(8)     ..TG..A................................A...................A...................
1b(6)     ..TG..A................................A...................A...................N........
```

Fig. 1 Five most dominant variants of HCV genotypes 3a and 1b present in a sample with mixed-genotype infection

genotype sample, whereas direct DNA sequencing has sensitivity of $\geq$25% (Caraballo Cortes et al. 2016). Consequently, a considerable number of samples with the undetermined, mixed, and unspecified subtype or even misclassified genotypes may occur. Due to the extensive output of sequence data, which enables the in-depth analysis of heterogeneous viral populations present in each sample, NGS may be considered a more reliable method (Minosse et al. 2016).

In the present study, we identified a mixed-genotype infection (3a + 1b) in one sample (1.3%) routinely diagnosed as harboring a single genotype (3a). The mixed infection with a lower percentage of one genotype in this patient could have resulted from a coinfection. Using NGS, two samples were determined as genotypes 1b and 4a, while a routine assay detected genotypes 1a and 3a, respectively. These discrepancies were probably caused by limited sensitivity of commercially used assays as none of them was based on the reference method analyzing HCV NS5B region, with subsequent phylogenetic analyses. A considerable proportion of genotype misclassified cases by commercially used tests, including a failure of differentiation between HCV subtypes 1a and 1b by Versant HCV Genotype 2.0 assay, have been recently described (Chueca et al. 2016). Such mistakes can lead to suboptimal antiviral treatment regimens affecting the patient response to therapy.

The distribution of HCV genotypes in a group of blood donors studied in the early phase of HCV infection of 1a, 5.1%; 1b, 42.4%; and 3a, 52.5% differs from the epidemiological data showing the predominance of genotype 1 in Poland's general population (79.4%) (Panasiuk et al. 2013) and in anti-HCV-positive donors (75.7%) and chronic hepatitis C patients (85.3%) (Grabarczyk et al. 2015). This inconsistency may be due to changes in the genotype-associated routes of HCV transmission and differences in the pathogenesis between particular viral genotype infections. As HCV transmission through medical procedures is considered to be low (Gowin et al. 2016), recently identified risk factors include the tattooing, injection and/or non-injection drugs, accidental exposure to blood, and sharing toiletries (Grabarczyk et al. 2015).

Mixed-genotype HCV infections have been found to be infrequent in the Polish population. Two studies including chronically infected individuals have demonstrated the prevalence of mixed-genotype infections at 1.6% (Panasiuk et al. 2013) and 2.2% (Gowin et al. 2016). These findings are in accord with the results of the present high-sensitivity study and with other studies performed in European countries, revealing that mixed HCV infections is rather an uncommon event. In the UK, prevalence was evaluated at 0.2% (Brant et al. 2010), and a lack of mixed HCV infections was reported in Sweden (Muhlberger et al. 2009).

In conclusion, this study performed on HCV-RNA-positive, anti-HCV-negative blood donors demonstrates that infection with more than one HCV genotype or subtype can occur during the very early stage of infection, supporting the possibility of simultaneous transmission of two different genotypes. Although not successful in all samples, next-generation sequencing enables the accurate identification of HCV mixed infections that are not detected by routine methods. Next-generation sequencing also helps rule out the inaccurate assignment of mixed infections by routine methods. Thus, higher resolution of next-generation sequencing enables the identification of minor variants of the viral population present in the sample as well as the clarification of ambiguous results.

Acknowledgments The study was supported by National Science Center grant No. 238240.

Conflicts of Interest The authors declare no conflicts of interest in relation to this article.

References

Altschul SF, Gish W, Miller W, Myers EW, Lipman DJ (1990) Basic local alignment search tool. J Mol Biol 215(3):403–410

Bagaglio S, Uberti-Foppa C, Di Serio C, Trentini F, Andolina A, Hasson H, Messina E, Merli M, Porrino L, Lazzarin A, Morsica G (2015) Dynamic of mixed HCV infection in plasma and PBMC of HIV/HCV patients under treatment with Peg-IFN/Ribavirin. Medicine (Baltimore) 94(43):e1876

Brant LJ, Ramsay ME, Tweed E, Hale A, Hurrelle M, Klapper P, Ngui SL, Sentinel Surveillance of Hepatitis Testing Group (2010) Planning for the healthcare burden of hepatitis C infection: hepatitis C genotypes identified in England, 2002-2007. J Clin Virol 48(2):115–119

Buoro S, Pizzighella S, Boschetto R, Pellizzari L, Cusan M, Bonaguro R, Mengoli C, Caudai C, Padula M, Egisto Valensin P, Palù G (1999) Typing of hepatitis C virus by a new method based on restriction fragment length polymorphism. Intervirology 42(1):1–8

Butt S, Idrees M, Ur Rehman I, Akbar H, Shahid M, Afzal S, Younas S, Amin I (2011) Mixed genotype infections with hepatitis C virus, Pakistan. Emerg Infect Dis 17(8):1565–1567

Caraballo Cortes K, Bukowska-Osko I, Pawelczyk A, Perlejewski K, Ploski R, Lechowicz U, Stawinski P, Demkow U, Laskus T, Radkowski M (2016) Next-generation sequencing of 5′ untranslated region of hepatitis C virus in search of minor viral variant in a patient who revealed new genotype while on antiviral treatment. Adv Exp Med Biol 885:11–23

Chueca N, Rivadulla I, Lovatti R, Reina G, Blanco A, Fernandez-Caballero JA, Cardenoso L, Rodriguez-Granjer J, Fernandez-Alonso M, Aguilera A, Alvarez M, Galan JC, Garcia F (2016) Using NS5B sequencing for hepatitis C virus genotyping reveals discordances with commercial platforms. PLoS One 11(4):e0153754

Giannini C, Giannelli F, Monti M, Careccia G, Marrocchi ME, Laffi G, Gentilini P, Zignego AL (1999) Prevalence of mixed infection by different hepatitis C virus genotypes in patients with hepatitis C virus-related chronic liver disease. J Lab Clin Med 134(1):68–73

Gowin E, Bereszynska I, Adamek A, Kowala-Piaskowska A, Mozer-Lisewska I, Wysocki J, Michalak M, Januszkiewicz-Lewandowska D (2016) The prevalence of mixed genotype infections in Polish patients with hepatitis C. Int J Infect Dis 43:13–16

Grabarczyk P, Kopacz A, Sulkowska E, Kubicka-Russel-D, Mikulska M, Brojer E, Letowska M (2015) Blood donors screening for blood born viruses in Poland. Przegl Epidemiol 69(3):473–477, 591–475

Hairul Aini H, Mustafa MI, Seman MR, Nasuruddin BA (2012) Mixed-genotypes infections with hepatitis C virus in hemodialysis subjects. Med J Malaysia 67(2):199–203

Hnatyszyn HJ (2005) Chronic hepatitis C and genotyping: the clinical significance of determining HCV genotypes. Antivir Ther 10(1):1–11

Hu YW, Balaskas E, Furione M, Yen PH, Kessler G, Scalia V, Chui L, Sher G (2000) Comparison and application of a novel genotyping method, semiautomated primer-specific and mispair extension analysis, and four other genotyping assays for detection of hepatitis C virus mixed-genotype infections. J Clin Microbiol 38(8):2807–2813

Huang CF, Huang CI, Yeh ML, Huang JF, Hsieh MY, Lin ZY, Chen SC, Yu ML, Dai CY, Chuang WL (2015) Host and virological characteristics of patients with hepatitis C virus mixed genotype 1 and 2 infection. Kaohsiung J Med Sci 31(5):271–277

Kao JH, Chen PJ, Lai MY, Yang PM, Sheu JC, Wang TH, Chen DS (1994) Mixed infections of hepatitis C virus as a factor in acute exacerbations of chronic type C hepatitis. J Infect Dis 170(5):1128–1133

Larkin MA, Blackshields G, Brown NP, Chenna R, McGettigan PA, McWilliam H, Valentin F, Wallace IM, Wilm A, Lopez R, Thompson JD, Gibson TJ, Higgins DG (2007) Clustal W and Clustal X version 2.0. Bioinformatics 23(21):2947–2948

Madalinski K, Zakrzewska K, Kolakowska A, Godzik P (2015) Epidemiology of HCV infection in Central and Eastern Europe. Przegl Epidemiol 69(3):459–464, 581–454

Martin M (2011) Cutadapt removes adapter sequences from high-throughput sequencing reads. EMBnet J 17(1):10

Minosse C, Giombini E, Bartolini B, Capobianchi MR, Garbuglia AR (2016) Ultra-deep sequencing characterization of HCV samples with equivocal typing results determined with a commercial assay. Int J Mol Sci 17(10):1679

Muhlberger N, Schwarzer R, Lettmeier B, Sroczynski G, Zeuzem S, Siebert U (2009) HCV-related burden of disease in Europe: a systematic assessment of incidence, prevalence, morbidity, and mortality. BMC Public Health 9:34

Panasiuk A, Flisiak R, Mozer-Lisewska I, et al (2013) Distribution of HCV genotypes in Poland. Przegl Epidemiol 67(1):11–16, 99–103

Parodi C, Culasso A, Aloisi N, Garcia G, Baston M, Corti M, Bianco RP, Campos R, Ares BR, Bare P (2008) Evidence of occult HCV genotypes in haemophilic individuals with unapparent HCV mixed infections. Haemophilia 14(4):816–822

Qian KP, Natov SN, Pereira BJ, Lau JY (2000) Hepatitis C virus mixed genotype infection in patients on haemodialysis. J Viral Hepat 7(2):153–160

Sakem B, Madalinski K, Nydegger U, Stepien M, Godzik P, Kolakowska A, Risch L, Risch M, Zakrzewska K, Rosinska M (2016) Hepatitis C virus epidemiology and prevention in Polish and Swiss population – similar and contrasting experiences. Ann Agric Environ Med 23(3):425–431

Schijman A, Colina R, Mukomolov S, Kalinina O, Garcia L, Broor S, Bhupatiraju AV, Karayiannis P, Khan B, Mogdasy C, Cristina J (2004) Comparison of hepatitis C viral loads in patients with or without coinfection with different genotypes. Clin Diagn Lab Immunol 11(2):433–435

Schroter M, Feucht HH, Zollner B, Schafer P, Laufs R (2003) Multiple infections with different HCV genotypes: prevalence and clinical impact. J Clin Virol 27(2):200–204

Widell A, Mansson S, Persson NH, Thysell H, Hermodsson S, Blohme I (1995) Hepatitis C superinfection in hepatitis C virus (HCV)-infected patients transplanted with an HCV-infected kidney. Transplantation 60(7):642–647

Zarkesh-Esfahani SH, Kardi MT, Edalati M (2010) Hepatitis C virus genotype frequency in Isfahan province of Iran: a descriptive cross-sectional study. Virol J 7:69

Adv Exp Med Biol - Clinical and Experimental Biomedicine (2018) 1: 73–82
https://doi.org/10.1007/5584_2018_197
© Springer International Publishing AG, part of Springer Nature 2018
Published online: 6 April 2018

Bioactive Oleic Derivatives of Dopamine: A Review of the Therapeutic Potential

Mieczyslaw Pokorski and Dominika Zajac

Abstract

Lipid derivatives of dopamine are a novel class of compounds raising a research interest due to the potential of their being a vehicle for dopamine delivery to the brain. The aim of the present paper is to review the main features of the two most prominent bioactive members of this family, namely, N-oleoyl-dopamine (OLDA) and 3′-O-methyl-N-oleoyl-dopamine (OMe-OLDA), with emphasis on the possible therapeutic properties.

Keywords

Brain · Dopamine receptor · Lipid derivatives of dopamine · Therapeutic potential · TRPV1 receptors

1 Lipid Derivatives of Dopamine: A New Class of Compounds

Lipid derivatives of dopamine, also described as N-acyl-dopamines and N-acyl-dopamides, are the products of condensation of dopamine and one of the fatty acids, preferably the oleic, arachidonic, stearic, or palmitic acid. Of the dopamide family, N-oleoyl-dopamine and N-arachidonyl-dopamine assume an active regulatory role in neural tissues,

with the stearic and palmitic derivatives being only enhancers of their activity (De Petrocellis et al. 2004; Chu et al. 2003). This review focuses on the two closely related and recently in-depth investigated dopamide compounds: N-oleoyl-dopamine (OLDA) and 3′-O-methyl-N-oleoyl-dopamine (OMe-OLDA) (Fig. 1a and b, respectively).

The history of oleic derivatives of dopamine begins with the presumption put forward by Pokorski and Matysiak (1998) pointing out the possibility of existence of bioactive lipid derivatives of biogenic amines, including dopamine. Soon afterward, the first method of de novo synthesis of these compounds has been described by Czarnocki et al. (1998), followed by reports on their bioactivity as TRPV1 agonists in vitro (Bisogno et al. 2000). In 2002, N-arachidonyl-dopamine has been found as a compound native to the mammalian brain tissue (Huang et al. 2002), followed shortly by N-oleoyl-dopamine (Chu et al. 2003). From that point on, extensive research on this group of compounds has begun. The research concentrated on the interaction of OLDA with the TRPV1 (vanilloid) receptor system. OLDA has been found to cause hyperalgesia and nocifensive behavior that are more intense than those after the archetype TRPV1 agonist capsaicin (Chu et al. 2003), which makes it one

M. Pokorski (✉)
Faculty of Physiotherapy, Opole Medical School, Opole, Poland
e-mail: pokorskim@wsm.opole.pl

D. Zajac
Laboratory of Respiration Physiology, Mossakowski Medical Research Centre PAS, Warsaw, Poland

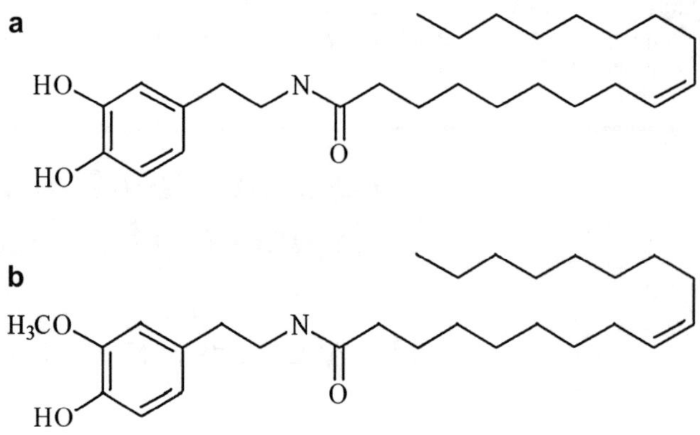

Fig. 1 Chemical structures
of (**a**) N-oleoyl-dopamine
(OLDA) and (**b**) 3′-O-
methyl-N-oleoyl-dopamine
(OMe-OLDA)

of the most active endovanilloids. However, the possibility of OLDA's interaction with the dopaminergic system has not been addressed at the time. Likewise, the bioactivity of 3′-O-methylated OLDA has remained unexplored, despite the plausibility that TRPV1 receptors show some preference in engaging compounds equipped with a methyl moiety. In addition, the exact physiological role of brain dopamides or their therapeutic potential has remained unsettled.

2 Chemical and Biological Synthesis and Metabolism of Oleic Derivatives of Dopamine

In general, there are two possible methods of chemical synthesis in vitro of OLDA and OMe-OLDA. The method proposed by Czarnocki et al. (1998) uses a peptide-like benzotriazol-1-yloxytris(dimethylamino)phosphonium hexafluorophosphate (BOP)-based coupling and gives a reaction yield of about 97%. The other, proposed by Bezuglov et al. (2001), the mixed anhydride-based method, gives a yield of up to 70%. In either, dopamine or 3′-O-methyl-dopamine is condensed with oleic acid or its derivative, respectively. Additionally, OMe-OLDA can be formed enzymatically from OLDA, using catechol-O-methyltransferase (COMT, EC 2.1.1.6) (Zajac et al. 2014).

N-acyl-dopamides may be synthesized in vivo from tyrosine in a dopamine-like manner (Hu et al. 2009). In case of OLDA, the first step is the acylation of tyrosine by oleic acid to N-oleoyl-tyrosine, followed by hydroxylation in position 3 on the phenyl ring by tyrosine hydroxylase (EC 1.14.16.2) and by decarboxylation by aromatic amino acid decarboxylase (EC 4.1.1.28) to N-oleoyl-dopamine.

OMe-OLDA is liable to be directly synthesized from OLDA by 3′O-methylation by COMT, the enzyme that metabolizes dopamine to 3-methoxytyramine. Zajac et al. (2014) have noticed that OLDA undergoes a transformation to OMe-OLDA in vitro with both commercially available COMT and with endogenous COMT natively present in the rat brain preparation, giving a reaction yield of about 3%. Exogenous OLDA, administered intra-arterially in the rat, also is methylated to OMe-OLDA with a yield of 9%; the reaction being is almost entirely blocked by tolcapone, a COMT inhibitor. This pathway is likely not only a synthetic pathway for OMe-OLDA but also the essential metabolic pathway of OLDA. OLDA itself is a poor ligand of the fatty acid amide hydroxylase (FAAH, EC 3.5.1.99) (Chu et al. 2003). Thus, a direct hydrolysis to dopamine and oleic acid is difficult to take place, making the classical dopamine-like metabolic pathway of oxidation by monoamine oxidase (MAO, EC 1.4.3.4) hardly, if at all, effective. Another possible pathway of elimination of OLDA is its sulfation. Akimov et al. (2009) have found that OLDA is in vitro sulfated by aryl sulfotransferase (EC 2.8.2.1) of liver and brain cells. Those authors argue that the sulfation process might be the most effective metabolic and eliminatory process for OLDA. However, this metabolic pathway is impossible for OMe-OLDA, as the sulfation takes place at the 3'hydroxyl group of dopamine moiety, occupied by the methyl group. Additionally, this pathway would involve some kind of competition between sulfation and methylation of OLDA which seems rather impossible under in vivo conditions. The activity of aryl sulfotransferases toward tyrosine and dopamine is only present in the cytosol (Fernando et al. 1993) where OLDA does not penetrate (Zajac 2009).

Results of the investigations hitherto performed on OLDA suggest that it is a relatively safe compound. OLDA does not induce any cytotoxicity or apoptosis in primary T cells (Sancho et al. 2004) and has no impact on the well-being of laboratory animals. Moreover, reserpine-treated rats seem to recover faster after experimental procedures performed with OLDA than the control healthy rats do (unpublished observation). As a compound per se, OLDA is stable both in vitro and in vivo. In the inorganic Krebs solution, OLDA remains unchanged for 4 h, which is comparable to dopamine. However, under oxidative conditions, OLDA is more stable than dopamine and decomposes after 2–4 h depending on the presence of calcium ions that hasten OLDA decomposition. In the rat brain membrane fraction, OLDA shows an outstanding stability for 17–24 h, also depending on the presence of calcium ions. The presence of membranes protects OLDA from oxidation (Przegalinski et al. 2006; Zajac et al. 2006). A possible explanation of OLDA stability is the micelle-like arrangement of membrane lipids, where dopamine moiety could be hidden inside and protected from oxidative milieu and calcium ions. In in vivo experiments, OLDA remains unchanged for up to 24 h after intraperitoneal injection (Zajac 2009).

3 Oleic Derivatives of Dopamine and the Brain

The physiological role of endogenous N-acyl-dopamines, especially of OLDA, is unknown.

Navarrete et al. (2010) have suggested that N-acyl-dopamines could be a kind of lipid neurotransmitters that stabilize cyclooxygenase-2 (COX-2), the effect hinging on the presence of the dopamine moiety. Since COX-2 is a protein that mediates the formation of prostaglandins from arachidonate and is engaged in proinflammatory prostanoid signaling, brain N-acyl-dopamines would play a neuroprotective role.

Chu et al. (2003) have found that OLDA is present in the bovine brain, although its exact concentration could not be settled. Huang et al. (2002) have unraveled a sister compound of OLDA, N-arachidonyl-dopamine (NADA), in the bovine striatum at a level of about 6 pmol/g wet tissue (wt), which corresponds to the concentration of 2.5 pg/mg wt. A closely comparable content of NADA has been noted by Ji et al. (2014) in the murine striata. Those authors have also assessed the content of OLDA at 0.15 pg/mg wt in the same matrix.

OLDA and OMe-OLDA are capable of penetrating neural tissues. Pokorski et al. (2006) have found that intra-arterially injected OLDA is absorbed by the carotid body, a peripheral blood barrier-free chemosensory organ of the neuroectodermic origin (Gonzalez et al. 1994), with a yield of about 30%, while it reaches the blood-brain-barrier equipped brain with a yield of 6%. In contradistinction, only 1.5% of OLDA, injected intravenously, is found in the brain (Pokorski et al. 2003). OLDA also is detectable in the brain after intraperitoneal administration, although the extent of that penetration has not been precisely verified (Zajac 2009). Likewise, intraperitoneally administered OMe-OLDA penetrates into the brain where it binds to brain membranes and remains stable for up to 24 h after injection (unpublished data). In addition, OLDA binds to rat brain membranes both in vitro and in vivo, where it remains stable for up to 24 h (Zajac et al. 2006). OLDA remains unchanged in frozen biological samples up to 2 weeks (Ji et al. 2014).

Hauer et al. (2013) have found endogenous OLDA in human plasma at a level of up to 0.45 ± 0.59 ng/ml, which suggests another, independent from the neural tissue, source of OLDA.

That may be not surprising as both dopamine and oleic acid are widely distributed in the body and body fluids. To date, the presence of endogenous OMe-OLDA has not been substantiated in the neural tissue, although it is highly likely as a consequence of OLDA metabolism.

4 Oleic Derivatives of Dopamine and TRPV1 Vanilloid Receptor System

Most of the research on lipid derivatives of dopamine has been focused on their interaction with TRPV1 vanilloid receptors. Both OLDA and OMe-OLDA show a strong interaction with the vanilloid system (Almasi et al. 2008). OLDA's binding to TRPV1 receptors, akin to other TRPV1 ligands, induces calcium influx into cells and, in turn, causes selective excitation of TRPV1-expressing neurons. OLDA injected intraplantarly in the rat decreases the latency of paw withdrawal from a heat source. It also produces nocifensive behaviors such as paw licking and lifting in reaction to painful heat stimuli; the hyperalgesia may last for as long as 3 h. These reactions are about 30-fold stronger and 4-fold longer than those observed after the use of the archetype vanilloid capsaicin. In addition, OLDA decreases the threshold for noxious heat. All these phenomena are inhibited by iodoresiniferatoxin (IRTX), a TRPV1 antagonist (Bolcskei et al. 2010; Szolcsanyi et al. 2004; Chu et al. 2003). However, OLDA's nociceptive behavior seems mediated not only by TRPV1 receptors but also via some other, by far undefined systems. Thermal hyperalgesia is also observed after intrathecal administration of OLDA (Spicarova and Palecek 2010). Interestingly, pretreatment with OLDA decreases the acute effect on nocifensive reactions of resiniferatoxin (RTX), a TRPV1 agonist. That raises the plausibility of a cross-desensitization between the two TRPV1 ligands (Bolcskei et al. 2010). Like other N-acyl-dopamines, OLDA is recognized by the anandamide transporter (Chu et al. 2003).

Another manifestation of OLDA's vanilloid-like properties is an increase in the frequency of the miniature excitatory postsynaptic currents (mEPSC) in the spinal dorsal horn (DH) neurons, engaged in transmission and modulation of peripheral nociceptive, pain-inducing messages as analyzed by a patch-clamp technique. A high concentration of OLDA (10 μM) is needed to evoke the mEPSC increase in the control condition. However, 50-fold lower concentration of OLDA suffices to increase the frequency mEPSC in the presence of PCK activators, bradykinin, or in animals with peripheral inflammation (Spicarova and Palecek 2009). Zhong and Wang (2008) have shown that OLDA protects the heart against ischemia-reperfusion injury via activation of TRPV1 receptors. The protection is mediated by increased release of substance P and calcitonin gene-related peptide (CGRP) as it is abrogated by blockade of CGRP receptors and K^+ channel antagonists. OLDA also hastens cardiac function recovery after ischemia-reperfusion episodes and decreases the release of lactate dehydrogenase, a marker of heart infarct. Yet another TRPV1-dependent feature of OLDA is its ability to decrease the viability of peripheral blood mononuclear cells, which along with the upregulation of TRPV1 may underlie immune dysfunction in chronic inflammation as exemplified by end-stage kidney disease (Saunders et al. 2009).

OLDA, like other vanilloids, when administered by the cerebroventricular route in the dose from 10 ng per rat up, aggravates the ictic indices and accelerates the development of seizures in both electrical and chemical models of epilepsy. In contradistinction, in the dose by an order of magnitude lower, 1 ng per rat or less, OLDA has the opposing anticonvulsant effect in the chemical model of epilepsy. In healthy rats, however, OLDA does not evoke epileptic seizures. The origin of this dichotomous pro- and anticonvulsant action is unknown. The knowledge is that the proconvulsant action of OLDA is mediated by TRPV1 receptors (Shirazi et al. 2014).

Butelman et al. (2004) have shown that OLDA produces allodynia after topical administration in primates akin to the effect of capsaicin. OLDA-induced allodynia is prevented by the antagonists of vanilloid receptors but also by pretreatment with the opioid receptor agonists loperamide and fentanyl. These observations point to the possibility of a close interplay between opioid and TRPV1 receptors in the pathogenesis of allodynia.

In contradistinction to OLDA, little is known about the interaction of OMe-OLDA with the TRPV1 receptor system. The biological plausibility exists that OMe-OLDA could have a more distinct vanilloid-like action due to its having the methyl moiety and carrying the structure-activity signaling, which is known to be of importance for TRPV1 function (Walpole et al. 1993a; Walpole et al. 1993b). It has been shown, however, that OMe-OLDA increases the intracellular calcium influx to a lower extent than capsaicin does. Nonetheless, OMe-OLDA produces thermal hyperalgesia and allodynia, in a typical vanilloid-like manner, with a decrease in heat pain threshold down to 10 °C (Almasi et al. 2008). The exact biochemical and signal transduction mechanisms of OMe-OLDA remain to be explored in alternative study designs.

5   Non-vanilloid and Non-dopaminergic Actions of OLDA

The actions of OLDA and OMe-OLDA are not restricted to the TRPV1 and dopamine receptor systems. Almaghrabi et al. (2014) have found that OLDA inhibits the ADP- and collagen-induced platelet aggregation in vitro in a non-vanilloid-like manner. This effect is related to the interaction of OLDA with the ADP-receptor system or to a change in platelet membrane properties after the compound incorporation. Another chemical structure-related action of OLDA is inhibition of anthrax lethal factor (ALF), a key virulence factor for *Bacillus anthracis*. Further, OLDA protects cells from the ALF-induced death being, by itself, non-cytotoxic. The action of OLDA relies on the two structural features: non-methylated catechol ring that enables the complexation of Zn^{2+} moiety of ALF and cis-double bond of the oleic acid tail (Gaddis et al. 2007). However, the latter's role is yet to be discerned (Gaddis et al. 2008).

An interesting feature of OLDA is the ability to improve glucose tolerance. Chu et al. (2010) have found that OLDA elevates the plasma level of the gastric inhibitory peptide, which induces insulin secretion via the G protein-coupled receptor 119 (GPR119) and improves time glucose tolerance in healthy animals. Hauer et al. (2013) have described that the level of circulating OLDA decreases three- to fourfold in traumatized patients who display symptoms of post-traumatic stress disorder and about twofold in traumatized patients without post-traumatic stress symptoms. In all patients, the levels of endocannabinoids and N-acyl-ethanolamides, but not OLDA, were higher than those in controls. Lack of the OLDA increase may be explained by the fact that, despite OLDA's structural resemblance to endocannabinoids, it is not a ligand for cannabinoid receptors (Chu et al. 2003). Visnyei et al. (2011) have found that OLDA inhibits a self-renewal of glioblastoma cells both in vitro and in vivo in a xenograft model. It not only inhibits the clonal sphere formation and cell proliferation but also reduces the sphere number. At the same time, OLDA has no effect on normal neural stem cells. OLDA also enhances release of dopamine and glutamate in striatal nerve terminals in a TRPV1-independent manner. That raises a specter of yet another, still undefined, mode of action of this class of substances, including some kind of modulation of neuronal transmission (Ferreira et al. 2009).

6   Antioxidant and Anti-inflammatory Properties of OLDA

OLDA interacts with a number of enzymes taking part in inflammation and oxidation. One of such enzymes is cyclooxygenase-2 (COX-2, EC 1.14.99.1) whose expression is upregulated by OLDA in brain endothelial cells, which, consequently, mitigates the propagation of an inflammatory reaction. This anti-inflammatory effect is neither dopamine- nor vanilloid-related and may be as a starting point for the activation of a natural barrier aimed at confining neuroinflammatory states (Navarrete et al. 2010).

OLDA also inhibits a series of lipoxygenases, notably arachidonate 5-lipoxygenase (Bezuglov et al. 2001) and 15-lipoxygenase that oxidizes linoleic acid (Nguyen et al. 2013). Bobrov et al. (2008) have shown that OLDA protects cultured cerebellar granule neurons from direct application of hydrogen peroxide, in contradistinction to dopamine proper (Daily et al. 1999). Thus, OLDA may be considered a natural, endogenous antioxidant. The antioxidant activity of OLDA is explicable by the impossibility of direct oxidation of OLDA by monoaminooxidase (MAO) due to blockade of the amine group (Bobrov et al. 2008). That observation has been confirmed by the authors of the present article who show that both OLDA and OMe-OLDA appreciably decrease the content of thiobarbituric acid reactive substances (TBARS) in the blood plasma of healthy animals. The decrease is from $0.694 \pm 0.023(SE)$ µmol/L in vehiculum-treated to 0.606 ± 0.042 µmol/L in OLDA-treated and to 0.664 ± 0.019 in OMe-OLDA-treated healthy animals ($p < 0.05$ both). The compounds also protect plasma from oxidative damage by hydrogen peroxide, decreasing the level of TBARS from 4.033 ± 0.433 µmol/L in the oxidized plasma to 0.623 ± 0.019 µmol/L and 0.696 ± 0.137 µmol/L in the OLDA and OMe-OLDA-pretreated plasma, respectively ($p < 0.01$ both). The decrease is on a par with the potent 0.616 ± 0.070 µmol/L decrease caused by vitamin E (unpublished data). Both compounds also decrease the content of lipid peroxidation products in plasma of reserpine-treated rats by about 25%, which is suggestive of a distinct antioxidant action (unpublished data). In addition, OLDA potently inhibits the activation events in the superantigen staphylococcal enterotoxin B-stimulated human peripheral T cells and prevents the entry of stimulated cells into the S phase of the cell cycle, showing immunosuppressant activity (Sancho et al. 2004).

Bioactivities of OLDA and OMe-OLDA above outlined raise the plausibility of potential therapeutic benefits of pharmacologic interventions changing the level of these native to the brain dopamides in pathologies involving central neural tissues.

7 Oleic Derivatives of Dopamine and the Dopaminergic System

Both OLDA and OMe-OLDA are clear enhancers of locomotor activity (Konieczny et al. 2009; Przegalinski et al. 2006) and suppressants of respiration (Zajac et al. 2018; Zajac 2009), the effects being reversed by dopamine antagonism. Despite the evident display of dopamine-like activity, the compounds are considered rather poor ligands of the dopaminergic receptor system as are also other N-acyl-dopamides (Bisogno et al. 2000). In the experiments on displacement of dopamine-radiolabeled agonists, OLDA and OMe-OLDA show a relatively low binding affinity to D2 dopamine receptors, with EC50 of 8.5 and 8.0 µM, respectively (unpublished data). Such high EC50 values indicate a limited interaction with the dopamine receptor system. However, dopamine proper also shows a relatively small binding capability, with the EC50 of 1 µM (Alachkar et al. 2010). A lower value of OLDA and OMe-OLDA binding to D2 dopamine receptors may be explicable by hindered access of OLDA and OMe-OLDA molecules to the binding site of dopamine receptor due to blockade of the amine group on dopamine ring by oleic acid (Missale et al. 1998).

In behavioral experiments, OLDA, at a dose of 10 mg/kg, significantly increases locomotor activity of freely moving rats from 403 ± 89 (SE) cm (DMSO-treated controls) to 1213 ± 196 cm (OLDA). This distinct threefold increase is abrogated by pretreatment with haloperidol, a D2 dopamine receptor antagonist, pointing to the interaction between OLDA and central D2 receptors (Przegalinski et al. 2006). Another feature of OLDA is its influence on passive extension and flexion of a rat hind limb at the ankle joint in the condition of enhanced muscle rigidity. OLDA in a dose of 20 mg/kg potently decreases the reserpine-enhanced tonic and reflex electromyographic activities both before and during movement and does that much faster and stronger than L-DOPA does. In reserpine-untreated rats, OLDA reduces muscle resistance even below the control level. However, OLDA does not influence catalepsy induced either by reserpine or by haloperidol (Konieczny et al. 2009). These observations indicate that OLDA exerts a muscle-relaxant action. Quite the same observations have been made concerning OMe-OLDA. It abolishes the effect on joint flexion and extension of the reserpine-induced enhancement of muscle rigidity much stronger and at lower doses than OLDA does, a decrease in maximum muscle torque of over 150 gcm for reserpine alone vs. about 100 gcm for 20 mg/kg OLDA after reserpine pretreatment vs. 75 gcm for 10 mg/kg OMe-OLDA after reserpine pretreatment. This myorelaxant activity lasts for about 1 h before it decreases toward the baseline level. Like OLDA, OMe-OLDA has no effect on reserpine- and haloperidol-induced catalepsy (unpublished data).

When another dopamine-dependent process, namely, regulation of respiration, is concerned, both compounds decrease normoxic and hypoxic ventilation in conscious freely moving rats. OLDA decreases normoxic ventilation by 29% and OMe-OLDA by 13%; both effects are mostly mediated by response in breathing frequency. Both compounds also attenuate the hypoxic ventilatory response, although the typical biphasic stimulatory/inhibitory response profile remains unaffected. Decreases in peak hypoxic ventilation amount 29% and 16% for OLDA and OMe-OLDA, respectively. A late ventilatory falloff brings lung ventilation down to about 60% of the baseline level in rats treated with both dopamides (Zajac et al. 2018). In anesthetized rats, OLDA decreases the hypoxic ventilatory response by about 20% (Zajac 2009). The interaction of OLDA with the respiratory regulation is liable to be mostly dopamine-dependent at the peripheral chemoreceptor level in contradistinction to OMe-OLDA that is presumed to act via both dopaminergic and TRPV1 vanilloid systems. The involvement of the latter system likely rests on a structural similarity of OMe-OLDA to the active center of TRPV1 receptors (Rekawek and Pokorski 2011). Stimulation by both dopamides of dopaminergic brain neurons via TRPV1 receptors located

on these neurons also is biologically plausible (Marinelli et al. 2003).

The understanding of the intertwined dopaminergic and non-dopaminergic TRPV1-linked interaction of OLDA and OMe-OLDA remains elusive. In case of hypoxic ventilatory regulation, such a double receptor target of dopamide compounds is difficult to reconcile with respiratory function. These compounds have an inhibitory peripheral effect on ventilation mediated by dopamine D2 receptors on carotid chemoreceptor cells. That closely resembles the inhibitory peripheral effect of dopamine, although it is stimulatory concerning ventilation at the central level (Gonzalez et al. 1994). The ventilatory role of TRPV1 receptors is less known. TRPV1 blockade with IRTX, a channel pore blocker, seems not to appreciably affect lung ventilation. Further, the inhibitory ventilatory effects of OLDA and OMe-OLDA administered on the background of TRPV1 blockade are sustained. These effects are counteracted by domperidone, a peripherally acting D2 receptor antagonist, which is congruent with the dopaminergic mediation at the carotid body level (Rekawek and Pokorski 2011; Zajac et al. 2010). Nonetheless, the issue of the ventilatory effects of TRPV1 receptors is by far unsettled and requires further exploration using alternative study designs that would enable the distinguishing between the involvement of peripheral and central receptor systems.

The research presented herein on oleic derivatives of dopamine is not only of scientific influence but has practical applications. Dopamides are model compounds for vanilloid research, carriers of dopamine molecule over the blood-brain barrier into the neural tissue, or even prodrugs of dopamine, all of which implies the potential therapeutic usefulness of dopamides in neurodegenerative conditions of brain tissue, particularly in the setting of dopamine insufficiency. The saga of unresolved physiologic mechanisms of dopamides' action and of the meaning of their presence in the mammalian brain goes on. Nonetheless, endogenous dopamides seem well suited to play a role of brain function modifiers. Exogenously administered dopamides, on the other hand, hold promise as therapeutic candidates for interaction with lipid signaling in homeostatic regulatory systems in the human body, such as lung ventilation. Pharmacological potential of OLDA and OMe-OLDA warrants further investigation and may be of interest for drug design research to help adjust the brain chemical imbalance.

Acknowledgments The dopamides reviewed in this article were synthetized in the Laboratory of Natural Products Chemistry, The Faculty of Chemistry, University of Warsaw. We remain appreciative of Dr. Zbigniew Czarnocki and Piotr Roszkowski's dexterity in the elaboration of de novo synthesis of the dopamides, and we are thankful for the provision of the compounds for our biological studies. We also thank Dr. Agnieszka Stasińska for giving a hand in some of the experiments described herein.

Conflicts of Interest The authors of this article are the inventors of the European (EP 2324824) and the US patents (US 8697898) covering the potential medical applications of OLDA and OMe-OLDA. The patent procedures were in part financed by the EU Innovative Economy grant POIG 1.3.2.-14–047/11.

References

Akimov MG, Nazimov IV, Gretskaya NM, Zinchenko GN, Bezuglov VV (2009) Sulfation of N-acyl dopamines in rat tissues. Biochem Mosc 74:681–685

Alachkar A, Brotchie JM, Jones OT (2010) Binding of dopamine and 3-methoxytyramine as L-DOPA metabolites to human alpha2-adrenergic and dopaminergic receptors. Neurosci Res 67:245–249

Almaghrabi SY, Geraghty DP, Ahuja KD, Adams MJ (2014) Vanilloid-like agents inhibit aggregation of human platelets. Thromb Res 134(2):412–417

Almasi R, Szoke E, Bolcskei K, Varga A, Riedl Z, Sandor Z, Szolcsanyi J, Petho G (2008) Actions of 3-methyl-N-oleoyldopamine, 4-methyl-N-oleoyldopamine and N-oleoylethanolamide on the rat TRPV1 receptor *in vitro* and *in vivo*. Life Sci 82:644–651

Bezuglov V, Bobrov M, Gretskaya N, Gonchar A, Zinchenko G, Melck D, Bisogno T, Di Marzo V, Kuklev D, Rossi JC, Vidal JP, Durand T (2001) Synthesis and biological evaluation of novel amides of polyunsaturated fatty acids with dopamine. Bioorg Med Chem Lett 11:447–449

Bisogno T, Melck D, Bobrov M, Gretskaya N, Bezuglov V, De Petrocellis L, Di Marzo L (2000) N-acyl-dopamines: novel synthetic CB1 cannabinoid-receptor ligands and inhibitors of anandamide inactivation with cannabimimetic activity in vitro and in vivo. Biochem J 351(Pt 3):817–824

Bobrov MY, Lizhin AA, Andrianova EL, Gretskaya NM, Frumkina LE, Khaspekov LG, Bezuglov VV (2008) Antioxidant and neuroprotective properties of N-arachidonoyldopamine. Neurosci Lett 431:6–11

Bolcskei K, Tékus V, Dézsi L, Szolcsanyi J, Petho G (2010) Antinociceptive desensitizing actions of TRPV1 receptor agonists capsaicin, resiniferatoxin and N-oleoyldopamine as measured by determination of the noxious heat and cold thresholds in the rat. Eur J Pain 14:480–486

Butelman ER, Harris TJ, Kreek MJ (2004) Antiallodynic effects of loperamide and fentanyl against topical capsaicin-induced allodynia in unanesthetized primates. J Pharmacol Exp Ther 311:155–163

Chu C, Huang S, De Petrocellis L, Bisogno T, Ewing S, Miller J, Zipkin R, Daddario N, Appendino G, Di Marzo V, Walker J (2003) N-oleoyldopamine, a novel endogenous capsaicin-like lipid that produces hyperalgesia. J Biol Chem 278:13633–13639

Chu ZL, Carroll C, Chen R, Alfonso J, Gutierrez V, He H, Lucman A, Xing C, Sebring K, Zhou J, Wagner B, Unett D, Jones RM, Behan DP, Leonard J (2010) N-oleoyldopamine enhances glucose homeostasis through the activation of GPR119. Mol Endocrinol 24:161–170

Czarnocki Z, Matuszewska I, Matuszewska M (1998) Highly efficient synthesis of fatty acids dopamides. Org Prep Proced Int 30:699–702

Daily D, Barzilai A, Offen D, Kamsler A, Melamed E, Ziv I (1999) The involvement of p53 in dopamine-induced apoptosis of cerebellar granule neurons and leukemic cells overexpressing p53. Cell Mol Neurobiol 19:261–276

De Petrocellis L, Chu CJ, Schiano Moriello A, Kellner JC, Walker JM, Di Marzo V (2004) Actions of two naturally occurring saturated N-acyldopamines on transient receptor potential vanilloid 1 (TRPV1) channels. Br J Pharm 143:251–256

Fernando PH, Sakakibara Y, Nakatsu S, Suiko M, Han JR, Liu MC (1993) Isolation and characterization of a novel microsomal membrane-bound phenol sulfotransferase from bovine liver. Biochem Mol Biol Int 30(3):433–441

Ferreira S, Lomaglio T, Avelino A, Cruz F, Oliveira C, Cunha R, Kofalvi A (2009) N-acyldopamines control striatal input terminals via novel ligand-gated cation channels. Neuropharmacology 56:676–683

Gaddis B, Avramova LV, Chmielewski J (2007) Inhibitors of anthrax lethal factor. Bioorg Med Chem Lett 17:4575–4578

Gaddis B, Rubert Perez C, Chmielewski J (2008) Inhibitors of anthrax lethal factor based upon N-oleoyldopamine. Bioorg Med Chem Lett 18:2467–2470

Gonzalez C, Almaraz L, Obeso A, Rigual R (1994) Carotid body chemoreceptors: from natural stimuli to sensory discharges. Physiol Rev 74(4):829–898

Hauer D, Schelling G, Gola H, Campolongo P, Morath J, Roozendaal B, Hamuni G, Karabatsiakis A, Atsak P,

Vogeser M, Kolassa IT (2013) Plasma concentrations of endocannabinoids and related primary fatty acid amides in patients with post-traumatic stress disorder. PLoS One 8(5):e62741

Hu SS, Bradshaw HB, Benton VM, Chen JS, Huang SM, Minassi A, Bisogno T, Masuda K, Tan B, Roskoski R Jr, Cravatt BF, Di Marzo V, Walker JM (2009) The biosynthesis of N-arachidonyl dopamine (NADA), a putative endocannabinoid and endovanilloid, via conjugation of arachidonic acid with dopamine. Prostaglandins Leukot Essent Fatty Acids 81:291–301

Huang S, Bisogno T, Trevisani M, Al-Hayani A, De Petrocellis L, Fezza F, Tognetto M, Petros T, Krey J, Chu C, Miller J, Davies S, Geppetti P, Walker J, Di Marzo V (2002) An endogenous capsaicin-like substance with high potency at recombinant and native vanilloid VR1 receptors. PNAS 99:8400–8404

Ji D, Jang CG, Lee S (2014) A sensitive and accurate quantitative method to determine N-arachidonoyldopamine and N-oleoyldopamine in the mouse striatum using column-switching LC–MS–MS: use of a surrogate matrix to quantify endogenous compounds. Anal Bioanal Chem 406:4491–4499

Konieczny J, Przegaliński E, Pokorski M (2009) N-oleoyl-dopamine decreases muscle rigidity induced by reserpine in rats. Int J Immunopathol Pharmacol 22:21–28

Marinelli S, Di Marzo V, Berretta N, Matias I, Maccarrone M, Bernardi G, Mercuri NB (2003) Presynaptic facilitation of glutamatergic synapses of the rat substantia nigra by endogenous stimulation of vanilloid receptors. J Neurosci 23:3136–3144

Missale C, Nash SR, Robinson SW, Jaber M, Caron MG (1998) Dopamine receptors: from structure to function. Physiol Rev 78(1):189–225

Navarrete CM, Pérez M, Garcia de Vinuesa A, Collado JA, Fiebich BL, Calzado MA, Muñoz E (2010) Endogenous N-acyl-dopamines induce COX-2 expression in brain endothelial cells by stabilizing mRNA through a p38 dependent pathway. Biochem Pharmacol 79:1805–1814

Nguyen MD, Nguyen DH, Yoo JM, Myung PK, Kim MR, Sok DE (2013) Effect of endocannabinoids on soybean lipoxygenase-1 activity. Bioorg Chem 49:24–32

Pokorski M, Matysiak Z (1998) Fatty acid acylation of dopamine in the carotid body. Med Hypotheses 50:131–133

Pokorski M, Matysiak Z, Marczak M, Ostrowski P, Kapuscinski A, Matuszewska I, Kanska M, Czarnocki Z (2003) Brain uptake of radiolabelled N-oleoyl-dopamine in the rat. Drug Dev Res 60:217–224

Pokorski M, Zajac D, Kapuscinski A, Matysiak Z, Czarnocki Z (2006) Accumulation of radiolabelled N-oleoyl-dopamine in the rat carotid body. Adv Exp Med Biol 580:173–178

Przegalinski E, Filip M, Zajac D, Pokorski M (2006) N-oleoyl-dopamine increases locomotor activity in the rat. Int J Immunopathol Pharmacol 19:897–904

Rekawek A, Pokorski M (2011) Influence of 3'-O-methyl-N-oleoyl-dopamine on the hypoxic ventilatory response in the rat. Conference on Advances in Pneumology, Bonn, Germany. http://www.pneumology.pl/bonn/media/doc/a089.pdf. Accessed on 13 Mar 2018 (Conference abstract)

Sancho R, Macho A, de La Vega L, Calzado MA, Fiebich BL, Appendino G, Munoz E (2004) Immunosuppressive activity of endovanilloids: N-arachidonoyl-dopamine inhibits activation of the NF-kB, NFAT, and activator protein 1 signaling pathways. J Immunol 172(4):2341–2351

Saunders CI, Fassett RG, Geraghty DP (2009) Up-regulation of TRPV1 in mononuclear cells of end-stage kidney disease patients increases susceptibility to N-arachidonoyl-dopamine (NADA)-induced cell death. Biochim Biophys Acta 1792:1019–1026

Shirazi M, Izadi M, Amin M, Rezvani ME, Roohbakhsh A, Shamsizadeh A (2014) Involvement of central TRPV1 receptors in pentylenetetrazole and amygdala-induced kindling in male rats. Neurol Sci 35:1235–1241

Spicarova D, Palecek J (2009) The role of the TRPV1 endogenous agonist N-oleoyldopamine in modulation of nociceptive signaling at the spinal cord level. J Neurophysiol 102:234–243

Spicarova D, Palecek J (2010) Tumor necrosis factor alpha sensitizes spinal cord TRPV1 receptors to the endogenous agonist N-oleoyldopamine. J Neuroinflamm 7:49

Szolcsanyi J, Sandora Z, Petho G, Varga A, Bolcskei K, Almasi R, Riedl Z, Hajos G, Czeh G (2004) Direct evidence for activation and desensitization of the capsaicin receptor by N-oleoyldopamine on TRPV1-transfected cell, line in gene deleted mice and in the rat. Neurosci Lett 361:155–158

Visnyei K, Onodera H, Damoiseaux R, Saigusa K, Petrosyan S, De Vries D, Ferrari D, Saxe J, Panosyan EH, Masterman-Smith M, Mottahedeh J, Bradley KA, Huang J, Sabatti C, Nakano I, Kornblum HI (2011) A molecular screening approach to identify and characterize inhibitors of glioblastoma stem cells. Mol Cancer Ther 10(10):1818–1828

Walpole CS, Wrigglesworth R, Bevan S, Campbell EA, Dray A, James IF, Perkins MN, Reid DJ, Winter J (1993a) Analogues of capsaicin with agonist activity as novel analgesic agents; structure-activity studies. 1. The aromatic 'A-region'. J Med Chem 36 (16):2362–2372

Walpole CS, Wrigglesworth R, Bevan S, Campbell EA, Dray A, James IF, Masdin KJ, Perkins MN, Winter J (1993b) Analogues of capsaicin with agonist activity as novel analgesic agents; structure-activity studies. 2. The amide bond 'B-region'. J Med Chem 36 (16):2373–2380

Zajac D (2009) Bioproperties of N-oleoyl-dopamine, a new lipid derivative of dopamine, with special attention to its influence on respiration in rats. PhD thesis, Mossakowski Medical Research Centre, Warsaw, Poland (in Polish)

Zajac D, Matysiak Z, Czarnocki Z, Pokorski M (2006) Membrane association of N-oleoyl-dopamine in the rat brain. J Physiol Pharmacol 57(Suppl 4):403–408

Zajac D, Roszkowski P, Czarnocki Z, Pokorski M (2010) Influence of TRPV1 (vanilloid) blockade on the respiratory response to hypoxia after N-oleoyl-dopamine in anesthetized rats. Conference on Advances in Pneumology, Warsaw, Poland. http://www.pneumology.pl/warsaw2010/media/doc/a059.pdf. Accessed on 13 Mar 2018 (Conference abstract)

Zajac D, Spolnik G, Roszkowski P, Danikiewicz W, Czarnocki Z, Pokorski M (2014) Metabolism of N-acylated-dopamine. PLoS One 9(1):e85259

Zajac D, Stasinska A, Pokorski M (2018) Oleic derivatives of dopamine and respiration. Adv Exp Med Biol 1023:37–46

Zhong B, Wang D (2008) N-oleoyl-dopamine, a novel endogenous capsaicin-like lipid, protects the heart against ischemia-reperfusion injury via activation of TRPV1. Am J Physiol Heart Circ Physiol 295:H728–H735

Adv Exp Med Biol - Clinical and Experimental Biomedicine (2018) 1: 83–91
https://doi.org/10.1007/5584_2018_189
© Springer International Publishing AG, part of Springer Nature 2018
Published online: 24 March 2018

Copeptin Blood Content as a Diagnostic Marker of Chronic Kidney Disease

Stanisław Niemczyk, Longin Niemczyk, Wawrzyniec Żmudzki, Marek Saracyn, Katarzyna Czarzasta, Katarzyna Szamotulska, and Agnieszka Cudnoch-Jędrzejewska

Abstract

Plasma content of copeptin increases with the advancement of chronic kidney disease (CKD). The purpose of this study was to evaluate copeptin content as a potential marker of CKD, as a single pathology or with coexisting heart failure. Seventy-six patients were divided into the following groups: Group 1 (control), without CKD and heart failure; Group 2, CKD stage 3a; Group 3, CKD stage 3b; Group 4, CKD stage 4; Group 5, CKD stage 5; and Group 6, CKD stage 3b and heart failure. For all patients, plasma concentrations of copeptin, creatinine, urea, cystatin C, sodium, C-reactive protein (CRP), N-terminal prohormone of brain natriuretic peptide (NT-proBNP), and blood pH were assessed. We found that plasma content of creatinine, urea, CRP, cystatin, NT-proBNP, and copeptin increased with CKD progression. Heart failure in CKD patients was not the cause of an appreciable increase of copeptin level. Copeptin/creatinine, copeptin/cystatin C ratios, and especially copeptin/eGFR ratio enhanced copeptin prognostic sensitivity concerning renal failure in CKD, compared with copeptin alone. The copeptin×NT-proBNP ratio decreased along CKD progression, reaching a nadir in the accompanying heart failure. In contradistinction, copeptin×NT-proBNP/creatinine ratio increased along CKD progression, reaching a peak in the accompanying heart failure. We conclude that copeptin is an important marker in CKD, but not so concerning heart failure in the disease. A decrease in copeptin×NT-proBNP and an increase in copeptin×NT-proBNP/creatinine ratio are useful markers of cardiac function decline in CKD.

S. Niemczyk, W. Żmudzki, and M. Saracyn
Department of Internal Medicine, Nephrology and Dialysis, Military Institute of Medicine, Warsaw, Poland

L. Niemczyk (✉)
Department of Internal Medicine, Nephrology and Dialysis, Military Institute of Medicine, Warsaw, Poland

Department of Nephrology, Dialysis and Internal Medicine, Medical University of Warsaw, Warsaw, Poland
e-mail: lniemczyk@wum.edu.pl

K. Czarzasta and A. Cudnoch-Jędrzejewska
Department of Experimental and Clinical Physiology, Laboratory of Center for Preclinical Research, Medical University of Warsaw, Warsaw, Poland

K. Szamotulska
Department of Epidemiology and Biostatistics, Institute of Mother and Child in Warsaw, Warsaw, Poland

Keywords
Biomarkers · Cardiac function · Chronic
kidney disease · Copeptin · Heart failure

1   Introduction

Chronic kidney disease (CKD) is a common
pathology for various age groups and is usually
a progressive disease, which frequently leads to
end-stage renal disease (ESRD) (Fogo 2006).
Diagnosing the condition, early and effective
treatment are important for mitigating its progression. Currently used biomarkers are not adequately sensitive and specific. Therefore, there
are problems with the assessment of advancement
and progression of CKD and with the confirma-
tion of kidney functions improvement (Tesch
2010; Lane et al. 2009).

Numerous studies suggest the usefulness of
evaluating vasopressinergic responses in severe
clinical conditions, including sepsis, pneumonia,
and stroke, but also in urinary tract and upper
respiratory tract infections (De Marchis et al.
2013; Boeck et al. 2012; Katan and Christ-Crain
2010; Jochberger et al. 2009; Morgenthaler et al.
2007). Arginine vasopressin (AVP) is an unstable
hormone, attached to platelets, and is practically
undetectable in plasma in clinical practice. For
that reason, copeptin, a C-terminal fragment of
pre-provasopressin, in plasma is used to assess
the activity of vasopressinergic responses
(Tasevska et al. 2016; Roussel et al. 2014).

Although the elevation of plasma copeptin
content in patients in critical conditions, such as
sepsis or severe pulmonary embolism, is
associated with an increase in the content of multiple stress hormones and cytokines, the diagnostic value of copeptin is still not fully determined,
including patients with CKD (Katan and Christ-
Crain 2010; Bhandari et al. 2009).

Roussel et al. (2014) have demonstrated that
copeptin content increases with progressing
CKD, which may suggest a causal relationship

of the two. An increase in copeptin content in
CKD is also confirmed by other studies that,
however, mostly concern the autosomal dominant
polycystic kidney disease, where copeptin content is proportional to both disease progression
and kidney size (Hu et al. 2015; Boertien et al.
2012).

Vasopressin regulates the water-electrolyte
balance and plasma osmolarity influencing renal
function. In case of impaired activity of vasopressinergic responses, heart failure can develop. An
attempt has been made to use copeptin as a
marker of heart failure severity, but the results
are equivocal (McMurray et al. 2012). Copeptin
has also been considered a marker cardiorenal
syndrome. Therefore, the purpose of this study
was to evaluate copeptin content as a potential
marker of CKD, as a single pathology or with
coexisting heart failure.

2   Methods

The study was approved by the Ethics Committee
of the Military Institute of Medicine in Warsaw,
Poland (permit 11/WIM/2013), and was
conducted in accordance with the Declaration of
Helsinki for Human Research of the World Medical Association. The study included 76 diabetes-
free patients, aged 38–76, with body mass index
(BMI) of 21–27 kg/m^2, suffering from polycystic
kidney disease and symptoms of infection. The
patients were treated with the renin-angiotensin-
aldosterone system (RAAS) inhibitors, known to
activate the vasopressinergic system. Patients
were divided into the following groups:

Group 1 (control group) – 11 healthy volunteers
    without CKD and without heart failure
    (estimated   glomerular   filtration   rate
    (eGFR)) $\geq$ 60 ml/min/1.73 m^2
Group 2–17 patients with CKD stage 3a (eGFR
    59–45 ml/min/1.73 m^2)
Group 3–12 patients with CKD stage 3b (eGFR
    44–30 ml/min/1.73 m^2)

Group 4–18 patients with CKD stage 4 (eGFR 29–15 ml/min/1.73 m^2)

Group 5–9 patients with CKD stage 5, in the predialysis period (eGFR <15 ml/min/1.73 m^2)

Group 6–9 patients with CKD stage 3b (eGFR 44–30 ml/min/1.73 m^2) and heart failure (ejection fraction <40% or NYHA class III or IV).

For all patients, concentrations of copeptin, creatinine, urea, cystatin C, sodium, potassium, calcium, C-reactive protein (CRP), N-terminal prohormone of brain natriuretic peptide (NT-proBNP) in the blood, blood and urine osmolarity, and blood pH were assessed once, using routine diagnostic tests. eGFR values were calculated using the MDRD formula (Levey et al. 2000). The values of copeptin/creatinine, copeptin/eGFR, copeptin/cystatin C, copeptin/N-terminal prohormone of brain natriuretic peptide (NT-proBNP), and copeptin×NT-proBNP/creatinine ratios were calculated.

Copeptin plasma concentration was measured using the enzyme-linked immunoassay (ELISA) (copeptin (human) EIA; EK-065-32; Phoenix Pharmaceuticals, Inc. Burlingame, CA).

All values presented in the text and tables are means ±SE or median and interquartile range (IQR). Nonparametric tests were used for descriptive statistical analysis. The Mann-Whitney U test was applied for comparisons between two groups, and the Jonckheere-Terpstra test was applied for trend analysis. For comparison of matched pairs, the exact Wilcoxon test was used. Where appropriate, chi-squared test, chi-squared test for trend, and McNemar's test were employed. A p-value < 0.05 defined statistical significance of differences.

3 Results

3.1 Characteristics of Patient Groups

The age of patients between groups was not significantly different: Group 1 (n = 11): 58.3 ± 7.5 vs. Group 2 (n = 17): 63.1 ± 10.7 vs. Group 3 (n = 12): 64.9 ± 8.3 vs. Group 4 (n = 18): 61.6 ± 11.7 vs. Group 5 (n = 9):

65.1 ± 5.3 years. However, Group 1 patients (58.3 ± 7.5) were significantly younger than Group 6 patients (n = 9) (65.4 ± 7.2) (p_{value} = 0.019). There were no statistically significant gender differences between the groups of patients with CKD ($p_{trend\ value}$ = 0.108).

3.2 Biochemical Indices

The eGFR value was statistically different between the groups of patients with CKD and healthy volunteers (Group 1) ($p_{trend\ value}$ < 0.001). There was a significantly lower eGFR value in Group 6 compared with Groups 1 and 2 patients (p_{value} < 0.001). Patients with CKD and heart failure patients (Group 6) had a significantly higher eGFR value compared with advanced CKD patients (Group 4, p_{value} = 0.029; Group 5, p_{value} ≤ 0.001). eGFR values in patients from Groups 6 and 3 were similar (Table 1).

Blood pH was significantly different in the studied groups (Group 1–Group 5) ($p_{trend\ value}$ = 0.013). It was higher in Group 5 compared with Group 6 patients (p_{value} = 0.029) (Table 1). Plasma sodium concentrations were not different between the groups (Table 1).

The plasma content of urea and creatinine was significantly different between the groups (Group 1–Group 5) ($p_{trend\ value}$ < 0.001). It was higher in Group 6 compared with Groups 1 and 2 patients (p_{value} < 0.001) and lower in Group 6 compared with Group 5 patients. (p_{value} = 0.001). There were no significant differences between the urea and creatinine content in Group 6 compared with Groups 3 and 4 patients. Plasma CRP was significantly different between the groups (Group 1–Group 5) ($p_{trend\ value}$ = 0.007). It was higher in Group 6 patients compared with healthy volunteers (Group 1) (p_{value} = 0.010). Cystatin C was significantly different between the groups of CKD patients and healthy volunteers (Group 1) ($p_{trend\ value}$ < 0.001). It was higher in Group 6 compared with Groups 1 and 2 patients (p_{value} < 0.001; p_{value} = 0.010, respectively) (Table 1).

The plasma content of NT-proBNP was significantly different between the groups (Group 1–

Table 1 Blood biochemical indices

Indices	Group 1 (n = 11)	Group 2 (n = 17)	Group 3 (n = 12)	Group 4 (n = 18)	Group 5 (n = 9)	Group 6 (n = 9)	
eGFR (ml/min/1.73m^2)	80.4 ± 14.2	52.9 ± 4.4	36.3 ± 4.3	23.3 ± 4.3	10.4 ± 2.8	30.4 ± 10.1	
	<0.001*	<0.001$^#$	0.173	0.029$^&$	<0.001$^@$		p_{value}
pH	7.35 ± 0.01	7.36 ± 0.03	7.36 ± 0.02	7.32 ± 0.07	7.31 ± 0.06	7.36 ± 0.01	
	0.231	0.968	0.612	0.103	0.029$^@$		p_{value}
Sodium (mmol/l)	139.1 ± 2.8	138.4 ± 1.7	138.3 ± 1.9	139.1 ± 2.4	135.4 ± 3.4	137.6 ± 2.5	
	0.233	0.466	0.512	0.209	0.209		p_{value}
Urea (mg/dl)	31.7 ± 12.0	50.5 ± 12.7	65.8 ± 19.4	107.4 ± 24.0	164.0 ± 34.6	96.0 ± 37.4	
	<0.001*	<0.001$^#$	0.072	0.471	0.004$^@$		p_{value}
Creatinine (mg/dl)	0.94 ± 0.22	1.32 ± 0.16	1.81 ± 0.27	2.92 ± 0.63	5.82 ± 1.39	2.54 ± 1.16	
	<0.001*	<0.001$^#$	0.113	0.065	0.001$^@$		p_{value}
†CRP (mg/l)	0.2 (0.1–0.3)	0.2 (0.1–0.6)	0.5 (0.2–0.8)	0.5 (0.1–1.1)	0.4 (0.2–1.6)	0.5 (0.3–1.8)	
	0.010*	0.063	0.282	0.407	0.681		p_{value}
Cystatin C (mg/l)	0.81 ± 0.21	1.34 ± 0.30	1.82 ± 0.30	2.73 ± 0.57	3.88 ± 0.81	2.64 ± 1.10	
	<0.001*	0.001$^#$	0.164	0.782	0.033$^@$		p_{value}
†NT-proBNP (pg/ml)	90.2 (33.9–149.7)	78.9 (49.2–673.8)	232.8 (119.0–309.7)	243.4 (80.7–464.0)	1128.6 (369.6–2620.0)	3003.0 (578.5–8020.0)	
	<0.001*	0.001$^#$	<0.001$^$$	0.001$^&$	0.139		p_{value}
†Copeptin (pmol/l)	17.55 (11.84–32.04)	25.77 (20.24–34.05)	27.28 (19.97–44.11)	33.03 (25.19–40.41)	41.13 (30.06–49.01)	39.69 (27.89–42.52)	
	0.002*	0.039$^#$	0.345	0.354	0.387		p_{value}

Data are means ±SE or medians (IQR), when marked (†), due to other than normal data distribution.
e-GFR, estimated glomerular filtration rate; Group 1, eGFR ≥60 ml/min/1.73m^2; Group 2, eGFR = 59–45 ml/min/1.73m^2; Group 3, eGFR = 44–30 ml/min/1.73m^2; Group 4, eGFR = 29–15 ml/min/1.73m^2; Group 5, eGFR <15 ml/min/1.73m^2; Group 6, ejection fraction (EF) < 40%; CRP, C-reactive protein; NT-proBNP, N-terminal prohormone of brain natriuretic peptide; *significant difference between Group 1 and Group 6; $^#$significant difference between Group 2 and Group 6; $^$$significant difference between Group 3 and Group 6; $^&$significant difference between Group 4 and Group 6; and $^@$significant difference between Group 5 and Group 6

Group 5) ($p_{trend\ value}$ < 0.001). It was higher in Group 6 patients compared with CKD only patients and compared with healthy volunteers. Copeptin also was significantly different between the groups (Group 1–Group 5) ($p_{trend\ value}$ < 0.001). It was higher in Group 6 compared with Group 2 patients and with healthy volunteers from Group 1 (p_{value} = 0.039; p_{value} = 0.002, respectively) (Table 1).

3.3 Ratio of Copeptin to Other Biochemical Indices

The copeptin/creatinine ratio was significantly different between the groups (Group 1–Group 5) ($p_{trend\ value}$ < 0.001). This ratio was greater in Group 6 patients with heart failure in comparison

with CKD patients in stage 5 (Group 5) (p_{value} = 0.001) (Table 2).

The copeptin/eGFR ratio differed significantly between the groups (Group 1–Group 5) ($p_{trend\ value}$ < 0.001). It was greater in Group 6 patients with heart failure in comparison with healthy volunteers (Group 1) (p_{value} < 0.001) and with CKD patients (Group 2, p_{value} < 0.001; Group 3, p_{value} = 0.049) but lower compared with CKD patients in stage 5 (p_{value} < 0.001). The copeptin/cystatin C ratio was significantly different between the groups (Group 1–Group 5) ($p_{trend\ value}$ < 0.001). There were no significant differences in the copeptin/cystatin C ratio between Group 6 patients and other groups of patients (Table 2).

The copeptin×NT-proBNP ratio was significantly different between the groups (Group 1–

Table 2 Ratio of plasma copeptin to other biochemical indices

Indices	Group 1 (n = 11)	Group 2 (n = 17)	Group 3 (n = 12)	Group 4 (n = 18)	Group 5 (n = 9)	Group 6 (n = 9)
Copeptin/creatinine	24.0 ± 16.3	21.5 ± 11.5	18.2 ± 14.1	12.1 ± 4.7	7.5 ± 2.9	16.4 ± 6.3
	0.503	0.241	0.808	0.067	**0.001**$^@$	p_{value}
Copeptin/eGFR	0.25	0.53	0.76	1.27	4.03	1.05
	(0.08–0.47)	(0.39–0.66)	(0.53–1.36)	(1.14–1.79)	(3.27–5.09)	(0.92–1.63)
	<0.001*	**<0.001**$^\#$	**0.049**$^\$$	0.348	**<0.001**$^@$	p_{value}
Copeptin/cystatin	27.5 ± 19.4	21.5 ± 12.3	18.3 ± 9.2	12.4 ± 3.3	10.6 ± 2.7	16.3 ± 7.3
	0.175	0.220	0.702	0.241	0.167	p_{value}
Copeptin/NT-proBNP	0.207	0.381	0.153	0.119	0.040	0.009
	(0.086–0.543)	(0.026–0.541)	(0.069–0.242)	(0.073–0.411)	(0.016–0.131)	(0.004–0.079)
	0.001*	**0.003**$^\#$	**0.001**$^\$$	**0.002**$^{\&}$	0.200	p_{value}
Copeptin x NT-proBNP/creatinine	1702.6	2384.9	2751.0	2460.8	7252.0	37759.1
	(943.8–3188.6)	(994.6–6161.0)	(1472.7–8573.7)	(1088.8–5926.8)	(2993.0–19639.7)	(13224.5–79962.1)
	<0.001*	**0.001**$^\#$	**<0.001**$^\$$	**<0.001**$^{\&}$	**0.011**$^@$	p_{value}

e-GFR, estimated glomerular filtration rate; Group 1, eGFR ≥ 60 ml/min/1.73m^2; Group 2, eGFR = 59–45 ml/min/1.73m^2; Group 3, eGFR = 44–30 ml/min/1.73m^2; Group 4, eGFR = 29–15 ml/min/1.73m^2; Group 5, eGFR <15 ml/min/1.73m^2; Group 6, EF < 40%; NT-proBNP, N-terminal prohormone of brain natriuretic peptide. *significant difference between Group 1 and Group 6; $^\#$significant difference between Group 2 and Group 6; $^\$$significant difference between Group 3 and Group 6; $^{\&}$significant difference between Group 4 and Group 6; $^@$significant difference between Group 5 and Group 6

Group 5) ($p_{\text{trend value}} = 0.024$). The value of the copeptin×NT-proBNP ratio was lower in Group 6 patients compared with healthy volunteers (Group 1) ($p_{\text{value}} = 0.001$) and with CKD patients (Group 2, $p_{\text{value}} = 0.003$; Group 3, $p_{\text{value}} = 0.001$; Group 4, $p_{\text{value}} = 0.002$) (Table 2).

Likewise, copeptin×NT-proBNP/creatinine ratio was significantly different between the groups (Group 1–Group 5) ($p_{\text{trend value}} = 0.025$). This ratio was greater in Group 6 patients with heart failure compared with healthy volunteers (Group 1) ($p_{\text{value}} < 0.001$) and with CKD patients (Group 2, $p_{\text{value}} = 0.001$; Group 3, $p_{\text{value}} < 0.001$; Group 4, $p_{\text{value}} < 0.001$). However, copeptin×NT-proBNP/creatinine ratio was significantly lower in Group 6 compared with Group 5 patients ($p_{\text{value}} = 0.011$) (Table 2).

4 Discussion

In the present study, there was a significantly increasing trend in plasma copeptin in patients with CKD progression. Yet copeptin was significantly greater in patients with CKD accompanied by heart failure compared with healthy volunteers. There were no significant changes in copeptin content between patients with CKD and heart failure compared with those without heart failure. It appears, therefore, that plasma copeptin content be a potential marker of chronic kidney disease.

That copeptin content increases with CKD progression has been shown, inter alia, by Tasevska et al. (2016) and Roussel et al. (2015), which raises the biological plausibility of a causal relationship of the two. Patients who are in need to start dialysis therapy have a greater higher copeptin content than those with no need to be dialyzed, despite similar creatinine clearance. The assessment of copeptin content is a predictor of CKD development in patients with diabetes (Tasevska et al. 2016; Boertien et al. 2012). A number of factors may cause copeptin to increase, such as elevated filtration pressure, renin–angiotensin–aldosterone system's vasospastic and procoagulant activity, inflammation, lipid disorders, and insulin resistance (Afsar 2017).

It appears that copeptin can be used as a diagnostic and prognostic marker of CKD because of its inverse correlation with eGFR (Engelbertz et al. 2016; Bhandari et al. 2009). In the present study, copeptin also tended to increase with a decrease in eGFR in CKD patients. This phenomenon can be explained by the fact that the plasma copeptin, a substance removed by kidneys, would increase with a worsening of kidney function. However, available studies show that eGFR is not a single determinant of copeptin content (Zittema et al. 2014). Ponte et al. (2015) have shown that an increase in copeptin serum concentration can occur before the eGFR starts to decrease.

Increased activity of the vasopressinergic system and release of copeptin into the circulation follow the development of CKD and worse urine densification (Zittema et al. 2012). Variability in plasma osmolarity has been shown to be a key factor influencing release of copeptin from the hypothalamus (Bankir et al. 2001). In the present study, there was a tendency for an increase in plasma content of urea, which may reflect an increase in plasma osmolarity (Hooper et al. 2015). The elevated urea content in plasma we noticed was accompanied by an increase in copeptin content. Other studies indicate that plasma copeptin level associates with both plasma and urine osmolarity, but the power of this association depends on eGFR (Roussel et al. 2014). Plischke et al. (2014) have shown that patients with CKD stages 1–4, with increased urine osmolarity, have a greater risk of progression to end-stage renal disease. Engelbertz et al. (2016) have found in a study carried out in CKD patients in stages 1–5 that copeptin plasma content increases with increasing creatinine content. Our present findings are in line with that observation. In addition, we also observed an association between the content of copeptin and cystatin C in CKD patients. The syntheses of creatinine and cystatin C are relatively constant, and the main route of their removal are kidneys. Therefore, creatinine and cystatin C plasma content reflects renal filtration of these agents and can be used to assess eGFR (Filler et al. 2005). Vuilleumier et al. (2016) have shown that cystatin C is of value as a prognostic marker of heart failure. We confirmed that in the present study, showing a significant decrease in plasma cystatin C content in CKD patients with heart failure.

In this study, we noticed an increase in plasma copeptin content accompanied by a trend in NT-proBNP elevation in CKD patients. Other studies have shown that an increase in plasma NT-proBNP content in CKD patients is a response of the myocardium to renal dysfunction (Niizuma et al. 2009). Yasuda et al. (2012) have shown that an increase in plasma NT-proBNP content associates with increased risk of accelerated progression of CKD to end-stage renal disease. However, Fenske et al. (2011) have found an increased incidence of cardiovascular events in patients with end-stage renal disease who have elevated plasma copeptin content. Engelbertz et al. (2016) have reported that increased copeptin plasma content could be a prognostic marker in patients with both coronary artery disease and CKD. Our present study demonstrates the usefulness of the copeptin×NT-proBNP/creatinine ratio as a good marker of heart failure in CKD. Further, we demonstrate that the copeptin×NT-proBNP ratio is also significant in this situation.

There are no similar reports in the literature that would evaluate the usefulness of the copeptin/creatinine, copeptin/eGFR, copeptin/cystatin, copeptin×NT-proBNP, or copeptin×NT-proBNP/creatinine ratios in various clinical situations. The findings of the present study suggest the importance of such calculated ratios in determining renal function and heart failure in CKD. The copeptin×NT-proBNP and copeptin×NT-proBNP/creatinine ratios can be very useful as markers for assessing heart failure in CKD patients. It seems, however, that the copeptin content by itself is of lower significance in heart failure, but it is quite a good and useful marker in CKD.

A limitation of this study is a small number of patients in each group studied and the lack of a group of patients with heart failure but without chronic kidney disease. However, patients with heart failure who would not be treated with RAAS are scarce, so that their recruitment is limited.

In conclusion, copeptin is an important marker in chronic kidney disease but not so concerning cardiac function in patients with chronic kidney disease. Copeptin/creatinine, copeptin/cystatin C, and copeptin/eGFR ratios may enhance copeptin prognostic sensitivity in chronic kidney disease. A decrease in copeptin×NT-proBNP and an increase in copeptin×NT-proBNP/creatinine ratio are useful markers for the assessment of heart failure in chronic kidney disease. Since there is a paucity of adequately sensitive and specific markers of kidney function in a chronic pathology, the search for new indicators is highly desirable to optimize clinical practice.

Acknowledgments Funded by a grant no. 307 from the Military Institute of Medicine. This research was carried out with the use of CePT infrastructure financed by the European Union Regional Development Fund within the Operational Program "Innovative Economy" for 2007–2013.

Conflicts of Interest The authors declare no conflicts of interest in relation to this article.

References

Afsar B (2017) Pathophysiology of copeptin in kidney disease and hypertension. Clin Hypertens 23:13

Bankir L, Bardoux P, Ahloulay M (2001) Vasopressin and diabetes mellitus. Nephron 87:8–18

Bhandari SS, Loke I, Davies JE, Squire IB, Struck J, Ng LL (2009) Gender and renal function influence plasma levels of copeptin in healthy individuals. Clin Sci (Lond) 116:257–263

Boeck L, Eggimann P, Smyrnios N, Pargger H, Thakkar N, Siegemund M, Morgenthaler NG, Rakic J, Tamm M, Stolz D (2012) The Sequential Organ Failure Assessment score and copeptin for predicting survival in ventilator-associated pneumonia. J Crit Care 27:523.e1–523.e9

Boertien WE, Meijer E, Zittema D, van Dijk MA, Rabelink TJ, Breuning MH, Struck J, Bakker SJ, Peters DJ, de Jong PE, Gansevoort RT (2012) Copeptin, a surrogate marker for vasopressin, is associated with kidney function decline in subjects with autosomal dominant polycystic kidney disease. Nephrol Dial Transplant 27:4131–4137

De Marchis GM, Katan M, Weck A, Fluri F, Foerch C, Findling O, Schuetz P, Buhl D, El-Koussy M, Gensicke H, Seiler M, Morgenthaler N, Mattle HP, Mueller B, Christ-Crain M, Arnold M (2013) Copeptin adds prognostic information after ischemic stroke: results from the CoRisk study. Neurology 80:1278–1286

Engelbertz C, Brand E, Fobker M, Fischer D, Pavenstädt H, Reinecke H (2016) Elevated copeptin is a prognostic factor for mortality even in patients with renal dysfunction. Int J Cardiol 221:327–332

Fenske W, Wanner C, Allolio B, Drechsler C, Blouin K, Lilienthal J, Krane V, German Diabetes, Dialysis Study Investigators (2011) Copeptin levels associate with cardiovascular events in patients with ESRD and type 2 diabetes mellitus. J Am Soc Nephrol 22:782–790

Filler G, Bökenkamp A, Hofmann W, Le Bricon T, Martínez-Brú C, Grubb A (2005) Cystatin C as a marker of GFR--history, indications, and future research. Clin Biochem 38:1–8

Fogo AB (2006) Progression versus regression of chronic kidney disease. Nephrol Dial Transplant 21:281–284

Hooper L, Abdelhamid A, Ali A, Bunn DK, Jennings A, John WG, Kerry S, Lindner G, Pfortmueller CA, Sjöstrand F, Walsh NP, Fairweather-Tait SJ, Potter JF, Hunter PR, Shepstone L (2015) Diagnostic accuracy of calculated serum osmolarity to predict dehydration in older people: adding value to pathology laboratory reports. BMJ Open 5:e008846

Hu W, Ni YJ, Ma L, Hao HR, Chen L, Yu WN (2015) Serum copeptin as a new biomarker in the early diagnosis of decline in renal function of type 2 diabetes mellitus patients. Int J Clin Exp Med 8:9730–9736

Jochberger S, Dörler J, Luckner G, Mayr VD, Wenzel V, Ulmer H, Morgenthaler NG, Hasibeder WR, Dünser MW (2009) The vasopressin and copeptin response to infection, severe sepsis, and septic shock. Crit Care Med 37:476–482

Katan M, Christ-Crain M (2010) The stress hormone copeptin: a new prognostic biomarker in acute illness. Swiss Med Wkly 140:w13101

Lane BR, Poggio ED, Herts BR, Novick AC, Campbell SC (2009) Renal function assessment in the era of chronic kidney disease: renewed emphasis on renal function centered patient care. J Urol 182:435–443

Levey AS, Greene T, Kusek JW, Beck GL, MDRD Study Group (2000) A simplified equation to predict glomerular filtration rate from serum creatinine (abstract). J Am Soc Nephrol 11:155A

McMurray JJ, Adamopoulos S, Anker SD, Auricchio A, Böhm M, Dickstein K, Falk V, Filippatos G, Fonseca C, Gomez-Sanchez MA, Jaarsma T, Køber L, Lip GY, Maggioni AP, Parkhomenko A, Pieske BM, Popescu BA, Rønnevik PK, Rutten FH, Schwitter J, Seferovic P, Stepinska J, Trindade PT, Voors AA, Zannad F, Zeiher A (2012) ESC Committee for Practice Guidelines. ESC guidelines for the diagnosis and treatment of acute and chronic heart failure

2012: the task force for the diagnosis and treatment of acute and chronic heart failure 2012 of the European society of cardiology. Developed in collaboration with the Heart Failure Association (HFA) of the ESC. Eur Heart J 33:1787–1847

Morgenthaler NG, Müller B, Struck J, Bergmann A, Redl H, Christ-Crain M (2007) Copeptin, a stable peptide of the arginine vasopressin precursor, is elevated in hemorrhagic and septic shock. Shock 28:219–226

Niizuma S, Iwanaga Y, Yahata T, Tamaki Y, Goto Y, Nakahama H, Miyazaki S (2009) Impact of left ventricular end-diastolic wall stress on plasma B-type natriuretic peptide in heart failure with chronic kidney disease and end-stage renal disease. Clin Chem 55:1347–1353

Plischke M, Kohl M, Bankir L, Shayganfar S, Handisurya A, Heinze G, Haas M (2014) Urine osmolarity and risk of dialysis initiation in a chronic kidney disease cohort–a possible titration target? PLoS One 9: e93226

Ponte B, Pruijm M, Ackermann D, Vuistiner P, Guessous I, Ehret G, Alwan H, Youhanna S, Paccaud F, Mohaupt M, Péchère-Bertschi A, Vogt B, Burnier M, Martin PY, Devuyst O, Bochud M (2015) Copeptin is associated with kidney length, renal function, and prevalence of simple cysts in a population-based study. J Am Soc Nephrol 6:1415–1425

Roussel R, Fezeu L, Marre M, Velho G, Fumeron F, Jungers P, Lantieri O, Balkau B, Bouby N, Bankir L, Bichet DG (2014) Comparison between copeptin and vasopressin in a population from the community and in people with chronic kidney disease. J Clin Endocrinol Metab 99:4656–4663

Roussel R, Matallah N, Bouby N, El Boustany R, Potier L, Fumeron F, Mohammedi K, Balkau B, Marre M, Bankir L, Velho G (2015) Plasma copeptin and decline in renal function in a cohort from the community: the prospective D.E.S.I.R. Study. Am J Nephrol 42:107–114

Tasevska I, Enhörning S, Christensson A, Persson M, Nilsson PM, Melander O (2016) Increased levels of copeptin, a surrogate marker of arginine vasopressin, are associated with an increased risk of chronic kidney disease in a general population. Am J Nephrol 44:22–28

Tesch GH (2010) Review: serum and urine biomarkers of kidney disease: a pathophysiological perspective. Nephrology (Carlton) 15:609–616

Vuilleumier N, Simona A, Méan M, Limacher A, Lescuyer P, E G, Bounameaux H, Aujesky D, Righini M (2016) Comparison of cardiac and non-cardiac biomarkers for risk stratification in elderly patients with non-massive pulmonary embolism. PLoS 11: e0155973

Yasuda K, Kimura T, Sasaki K, Obi Y, Iio K, Yamato M, Rakugi H, Isaka Y, Hayashi T (2012) Plasma B-type natriuretic peptide level predicts kidney prognosis in patients with predialysis chronic kidney disease. Nephrol Dial Transplant 27:3885–3891

Zittema D, Boertien WE, van Beek AP, Dullaart RP, Franssen CF, de Jong PE, Meijer E, Gansevoort RT (2012) Vasopressin, copeptin, and renal concentrating capacity in patients with autosomal dominant polycystic kidney disease without renal impairment. Clin J Am Soc Nephrol 7:906–913

Zittema D, van den Berg E, Meijer E, Boertien WE, Muller Kobold AC, Franssen CF, de Jong PE, Bakker SJ, Navis G, Gansevoort RT (2014) Kidney function and plasma copeptin levels in healthy kidney donors and autosomal dominant polycystic kidney disease patients. Clin J Am Soc Nephrol 9:1553–1562

Adv Exp Med Biol - Clinical and Experimental Biomedicine (2018) 1: 93–103
https://doi.org/10.1007/5584_2018_191
© Springer International Publishing AG, part of Springer Nature 2018
Published online: 30 March 2018

Psychological Determinants of Attitude Toward Euthanasia: A Comparative Study of Female Nurses and Female Nonmedical Professionals

Alicja Głębocka

Abstract

Moral, legal, and psychological aspects of the legality of euthanasia are subject to debates and studies of various communities. Diagnosing attitudes toward euthanasia should involve not only determining the proportion between its supporters and opponents but also the describing of mechanisms behind the development of particular views. The aim of the present study was to determine the psychological determinants of attitudes, such as fear of death-dying, self-esteem, and mood. The methods consisted of using the following questionnaires: the Głębocka-Gawor Attitudes Toward Euthanasia Inventory, the Ochsmann Fear of Death and Dying Inventory, the Dymkowski Self-Description Scale, the Adamczyk-Glebocka Negative Mood Inventory, and a measure of unconscious fear of death. The study involved 49 female nurses and 43 female nonmedical professionals. The results demonstrate that the attitudes and fear of death-dying did not differentiate the two groups of participants. Although the fear of dying weakened the strength of conservative views, it also reinforced the need for informational and psychological support. A high self-esteem was a predictor of conservative attitudes, while negative mood predicted liberal attitudes. Conservative attitudes were connected to a hidden fear of death and high self-esteem, while liberal attitudes were linked to a conscious fear and a rational vision of the self, the world, and the future.

Keywords

Conservative attitudes · Euthanasia · Fear of dying · Mood · Liberal attitudes · Nurses · Self-esteem

1 Introduction

1.1 Euthanasia

Euthanasia is defined as a deliberate action to end the life of a terminally ill patient by administering a lethal drug. There is a distinction between active euthanasia and passive euthanasia, which involves discontinuation of treatment or withdrawal of life-sustaining care leading to the death of a patient (Kirmes et al. 2017; Gesang 2008). There is also a third way to end the life of terminally ill patients, which is physician-assisted suicide. In this case, patients themselves administer a lethal substance prepared by a physician.

Supporters of euthanasia point out that patients facing inhuman and unbearable suffering should

A. Głębocka (✉)
The Andrzej Frycz Modrzewski University, Cracow, Poland
e-mail: aglebocka@afm.edu.pl

be given an opportunity to die with dignity. Euthanasia opponents hold a view that human life is the greatest value and cannot be terminated because of pain or suffering, but these should be alleviated with all means offered by modern medicine. Furthermore, the opponents claim that legalization of euthanasia opens the door to abuses, such as killing the mentally ill or elderly or patients whose care has become a burden on their relatives. In response to these objections, the supporters argue that euthanasia procedures provide safeguards that nobody will be killed insidiously, hastily, or, last but not least, against their will (Głębocka et al. 2013).

In numerous countries, including Poland, legal regulations prohibit euthanasia (Kirmes et al. 2017). However, with growing public acceptance of this practice, the number of countries which have legalized all or selected forms of euthanasia (active, passive, or assisted suicide) has been gradually increasing (Danyliv and O'Neill 2015; Cohen et al. 2013). Recent years have seen legalization of assisted suicide in Germany and Canada, as well as the right to active euthanasia for children in Belgium. There is a discussion on the acceptability of euthanasia for the mentally ill (Davidson and Lymburner 2017).

Various determinants of attitudes toward euthanasia have been described in the literature on the subject. These include religious beliefs, conservative worldview, age, admissibility of euthanasia in a given country, and medical profession. Study results indicate that euthanasia is accepted, sometimes under certain conditions, by people who are low to moderately religious, young, living in countries that allow euthanasia, and practicing a medical profession (Roelands et al. 2015; Gielen et al. 2009; Ho and Penney 1992).

All the aforementioned factors determining attitudes toward euthanasia are of sociocultural nature. Only a few studies address psychological dimensions, such as locus of control or extraversion, correlating with negative attitudes toward euthanasia (Aghababaei et al. 2014; Hains and Hulbert-Williams 2013). In addition to seemingly cross-cultural psychological predictors of attitudes toward euthanasia, particularly honesty,

humility, and openness to experience, a number of culturally relevant personality traits can be identified, such as agreeableness, which plays a major role among Iranians. Also altruism has been shown to be significantly correlated with nonacceptance of euthanasia (Aghababaei 2014; Aghababaei et al. 2014). Since attitudes toward euthanasia are determined by personality factors, and a desire for quick death among terminally ill patients is connected with anxiety and depression, a question arises if and how attitudes toward euthanasia are related to a specific type of fear, i.e., fear of death.

1.2   Fear of Death and Fear of Dying

Fear of death is the most universal, common, and inevitable human experience. Its origin stems from the imaginative abilities which make individuals create unverifiable visions of death. This type of anxiety appears to outweigh other fears, though it is often not as direct as the latter are, as it is not based on individual experience but rather on one's ideas of what may happen. Indirect death experience, always with respect to other people, makes death an underdefined, mysterious, and irreversible phenomenon, hence raising anxiety. Death anxiety can also stem from loneliness or alienation not only from one's social environment but also toward oneself. Fear of death is rooted in two basic instincts: the preservation of the self and the preservation of the species. According to the terror management theory (TMT), fear of death does not leave people indifferent. In order to cope with death-related anxiety, they employ proximal (direct) and distal (symbolic) defenses. Proximal defenses involve conscious, rational attempts to push thoughts of death out of focal attention (Pyszczynski et al. 1999), whereas distal defenses implement anxiety buffers by sustaining beliefs in the validity of the adopted worldview as well as maintaining and improving self-esteem. The worldview is defined as a collective system of meanings, consisting of cherished values, rules of conduct, customs, ceremonies, rituals, traditions, habits, as well as political and religious views considered valid.

This collective system of meanings sets the boundaries of norms and deviations, and is a point of reference for the assessment of behaviors, offering the explanation of phenomena occurring in the world. This enables an individual to effectively reduce uncertainty and anxiety while increasing the sense of security. One's worldview provides protection only if certain conditions are met, namely, an individual must believe that he has a valid and important worldview and identifies with it. In addition, an individual must have high self-esteem and the sense of being capable of living up to the standards of his worldview (Rosenblatt et al. 1989). The more common a particular worldview, the more effective it is in reducing fear of death (Arndt et al. 1997). As for anxiety coping mechanisms related to self-esteem, it is assumed that high self-esteem provides protection from fear and modifies the accessibility of death-related thoughts. When faced with death-thought accessibility, an individual can effectively cope with anxiety by boosting his self-esteem (Śmieja et al. 2006). According to contemporary scholars, the mechanisms of worldview defense and self-esteem manipulation described by TMT (Greenberg et al. 1992) need to be supplemented by mood regulation mechanisms, which effectively reduces death anxiety (Kneer and Rieger 2016).

Fear of death is often equated by investigators with fear of dying. However, even though both dimensions are moderately correlated, they should be considered separately, as rightly pointed out by Wittkowski (2001) who has extended the Collett and Lester (1969) approach of looking into death acceptance. All the four dimensions specified by him, i.e., fear of one's own and another person's death and dying, are relatively independent of one another as well as stable, i.e., not susceptible to changes over time or due to situational stimuli. Ochsmann (1984) has proposed even more detailed division of fear of death and dying. This concept encompasses six main dimensions: fear of encountering death, i.e.,

anxiety about having direct contact with a dying or dead person; fear of mortality, connected with anxiety about one's plans and expectations that will be shattered by death; fear of the end of one's life, expressed in not accepting death as a definite end of existence; fear of physical destruction, connected with a strong concern about what will happen with one's body after death; fear of life after death, evoked by uncertainty about what happens after death; and fear of dying, connected with one's image of suffering that dying involves.

1.3   Research Hypotheses

The aim of the present study was to determine the relationship between attitudes toward euthanasia and selected psychological dimensions, such as fear of death and fear of dying, death-thought accessibility, negative mood, and self-esteem.

The following research hypotheses have been formulated:

– Nurses will demonstrate lower magnitude of fear of death and dying than lay women in the control group. It seems self-evident that suppressing fear of death and dying is a major precondition of practicing any medical profession. Otherwise, health-care professionals would incur high psychological costs, while the anxiety experienced by them would affect the quality of their work. Dehumanization practices might, to an extent, play a protective role in death anxiety management (Vaes and Muratore 2013; Haque and Waytz 2012).
– Nurses will be more supportive of the admissibility of euthanasia and more appreciative of informational support than the control group subjects. Greater acceptance of euthanasia among health-care professionals has been demonstrated in a number of studies. Hence, it may be assumed that medical knowledge and professional experience are of significant importance in this respect.

– Fear of dying will be predictive of attitudes permitting euthanasia. Fear of dying is understood as anxiety about the process of dying, which involves great psychological and physical suffering. The willingness to avoid such suffering is indicated by a lot of patients asking for a judicial consent for euthanasia as the key motive behind their decision on early life termination (Głębocka et al. 2013).

– Fear of death will be predictive of attitudes forbidding euthanasia and requesting to sustain life as the highest value. Fear of the unknown, i.e., of what happens to a human being after death, may increase the need to avoid confrontation with such experience and to sustain life regardless of the psychological costs borne by a patient.

– Death-thought accessibility will intensify attitudes questioning the admissibility of euthanasia. Most likely, unconscious fear of death may contribute to avoiding confrontation with death for as long as possible. Similar mechanisms have been described in case of other unconscious fears, such as phobias, in which patients consistently try to avoid the subject of their fear, or conversion disorders, which involve unconscious avoidance of stressful situations through neurological dysfunctions, such as paralysis (Butcher et al. 2016).

– Negative mood (emotions and motivations) will intensify attitudes accepting euthanasia. Experiencing the feelings of sadness, hopelessness, or despair may contribute to a negative perception of not only oneself and the world around but also the future.

– High self-esteem will be predictive of negative attitudes toward the admissibility of euthanasia. Psychologists agree that high self-esteem is shared by most people. In Polish studies, over 80% of people have high self-esteem. Owing to its overt nature, it may be suspected that this rather realizes the need for social acceptance. The latter is manifested twofold: as a tendency to present oneself in a positive light and a tendency to meet expectations of others. As demonstrated by several studies, there is a negative attitude toward euthanasia in the Polish society, which may be traced back to religious beliefs.

2 Methods

This project received a positive recommendation from the Research Committee of the Faculty of Psychology and Humanities in the Andrzej Frycz Modrzewski Cracow University. The study involved 92 women, including 49 nurses working in outpatient clinics or nonsurgical departments and a control group of 43 nonmedical professionals. The mean age was 43 ± 10 (SD) years. Participation in the survey was voluntary and completely anonymous. The study was conducted in the Polish cities of Cracow and Opole. In each case, the respondents were reached individually in their workplace.

2.1 Survey Questionnaires

The following survey methods were applied in the study:

– The Attitudes Toward Euthanasia Inventory was used to determine the acceptance of euthanasia (Głębocka et al. 2013). The questionnaire consists of 28 items over three scales: (1) informational support scale (12 items), which measures the need of patients and their relatives for information and psychological support from physicians and nurses; (2) liberal attitude scale (9 items), which includes statements on the admissibility of euthanasia and the right of terminally ill patients to decide on early termination of life; and (3) conservative attitude scale (7 items), which consists of statements reflecting a conviction about absolute, unconditional nonacceptability of euthanasia owing to the value of human life and lack of moral authority to decide on life termination. Participants were to indicate to what extent they agreed with particular statements on a five-point scale from 1 ("Definitely no") to 5 ("Definitely yes"). The questionnaire reliability is acceptable (Cronbach's alpha = 0.83).

- The Fear of Death and Dying Inventory by Ochsmann (1984) was applied to measure the declared death and dying anxiety. Four out of six scales of the questionnaire were used in the study. These included (1) fear of encountering death, i.e., anxiety about having direct contact with a dying or dead person; (2) fear of mortality, involving anxiety evoked by a thought that one's plans would not be completed before death and that death would prevent care for close relatives and cause them pain and suffering; (3) fear of the end of one's life, involving anxiety that every life has an end and one would have to die someday, as well as sadness that one is not immortal; and (4) fear of dying, involving anxiety about dying accompanied by suffering, slow death, and long and painful illness. Participants were to answer questions on a five-point scale from 1 ("Strongly disagree") to 5 ("Strongly agree") The reliability of individual scales is acceptable (Cronbach's alpha from 0.87 to 0.97).

- Self-esteem was assessed using the self-description scale by Dymkowski (1993), which measures self-esteem in 14 selected dimensions, such as independent thinking, self-control in difficult situations, optimism, and cheerfulness. Participants were to rate each trait in themselves on a scale of −5 to +5. The questionnaire reliability is acceptable (Cronbach's alpha = 0.92).

- The current affective state was investigated using the Negative Mood Inventory (NMI) by Klaudia Adamczyk and Alicja Głębocka (unpublished inventory), which consists of two scales: (1) negative emotions and motivations (26 statements), e.g., *I feel sad, I am dwelling on my own failures, I feel insecure*, and *I don't feel like doing anything*, and (2) negative thoughts about oneself

(9 statements), e.g., *I am worthless, only misfortunes happen to me*, and *I am a looser*. The participating women rated on a scale of 1–5 the extent to which each statement was true of their feelings. The validity of the method has been confirmed in comparative studies of patients with depressive disorders vs. healthy control subjects. In addition, strong correlations between NMI scales and the UWIST Mood Adjective Checklist (UMACL, Matthews et al. 1990) in the Polish adaptation by (Goryńska 2005) have been demonstrated. The questionnaire reliability is acceptable (Cronbach's alpha = 0.97).

- Death-thought accessibility and unconscious fear of death were measured using projection methods for investigating the unconscious fear of death. These included (1) association scale, in which participants were to make phrases with 14 specific words, e.g., *enemy, silence*, and *kiss* (for each association with death, such as *kiss of death*, they scored one point); (2) image, in which participants, having been shown the drawing *All is Vanity* by Charles Allan Gilbert, were to quickly specify what they saw in the picture: a woman in front of a mirror or a human skull (if they saw a skull first, it meant higher unconscious fear of death); and (3) sentence – in this task, participants were to finish the sentence *What I am most afraid of is* .... The answers were categorized into those directly or indirectly associated with death, e.g., *What I am most afraid of is that I will have cancer*, and other, e.g., *What I am most afraid of is that I will have no job*.

2.2 Statistical Elaboration

Data were presented as means ±SD. The scores on the scales measuring conscious fear of death were subjected to multivariate analysis of variance (MANOVA) for repeated measures with profession being the group factor (nurses vs. nonmedical professionals, i.e., lay women in the control group). Scheffe's post hoc test was used to examine differences between four dimensions in each factor. Multiple regression analysis was conducted to define the predictive value on the attitudes toward euthanasia of various dimensions of death anxiety in the entire cohort combining both groups of women. A p-value <0.05 defined statistically significant changes. A commercial statistical package of Statistica v14 software was used for data analysis (StatSoft; Tulsa, OK).

3 Results

Significant differences were found between nurses and lay women. Nurses had significantly lower mean scores than lay women, 3.2 ± 0.7 vs. 3.5 ± 0.8, respectively ($F(1.90) = 4.59$; $p = 0.03$, $\eta^2 = 0.06$), which means the nurses declared lower death anxiety. Scheffe's *post hoc* test did not indicate any significant differences between the four dimensions considered in each group, namely, fear of encountering death, fear of mortality, fear of the end of one's life, and fear of dying.

Comparisons of dependent variables in the entire cohort ($F(3.27) = 15.39$, $p = 0.00001$, $\eta^2 = 0.18$) revealed that women, regardless of their workplace, were most afraid of dying (mean score 3.7 ± 0.7), followed by being mortal (3.4 ± 0.7), encountering death (3.3 ± 0.8), and end of one's life (3.0 ± 1.0). There were significant differences between fear of dying and the other dimensions ($p < 0.001$) and between fear of end of one's life and fear of mortality ($p < 0.05$).

Moreover, significant within-group interactions were present, which indicate that nurses were less afraid of encountering death and dying than of being mortal themselves, whereas lay women in the control group were, vice versa, more afraid of encountering death and dying and less afraid of their own mortality ($F(3.27 = 2.72$; $p = 0.04$) (Fig. 1).

To determine intergroup differences in attitudes toward euthanasia and in other dimensions, MANOVA analysis was conducted against the group factor of workplace (nurses vs. lay controls). The detailed results are presented in Table 1. The results show that no intergroup differences were found for most variables. Nurses did not differ from lay women in the control group in terms of attitudes toward euthanasia, namely, informational support, liberal attitudes, and conservative attitudes ($F(1.92) = 0.63$; $p > 0.05$), as well as self-esteem ($F(1.92) = 0.23$; $p > 0.05$) and negative affect (Wilks' lambda = 0.99, $F(2.92) = 0.16$; $p > 0.05$). Further, as MANOVA conditions were not satisfied, nonparametric Mann-Whitney U test was applied to the association scale. The test failed to demonstrate any significant differences between the groups either ($U = 512.5$, $Z = -0.3$; $p > 0.05$).

Death-thought accessibility (DTA) was measured using not only the association scale but also two projection methods: incomplete sentence and double-meaning images. The sentence "What I am most afraid of is ..." was finished in a way indicating death by 21.9% nurses and 28.6% lay women in the control group ($p > 0.05$). Concerning the image projection, the share of those seeing a skull first also was similar in both groups, namely, 63.7% for nurses vs. 58.3% for controls ($p > 0.05$).

As there were no significant differences between the two groups in terms of fear of death and attitudes toward euthanasia or other variables, namely, self-esteem, negative affect, and unconscious fear of death, it was assumed that the female participants were a homogeneous group. Consequently, further analysis was carried out for the entire cohort. Multiple regression analysis was conducted to define the predictive effect on the attitudes toward euthanasia of variables measuring various dimensions of death anxiety. Only fear of dying was found to be a significant predictor of informational support (unstandardized regression coefficient B = 0.34, $t(92) = 2.33$; $p < 0.05$) and of nonacceptance attitudes toward euthanasia (B = -0.34, $t(92) = -2.40$; $p < 0.05$). It means that fear of dying enhances the need for informational and psychological support and reduces a conviction that life needs to be supported unconditionally.

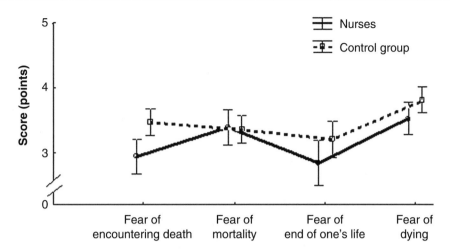

Fig. 1 Death anxiety dimensions; intergroup comparison

Table 1 Intergroup comparisons in attitudes toward euthanasia

Dimensions – dependent variables	Nurses	Controls	p
Informational support	4.3 ± 0.6	4.2 ± 0.6	0.49
Liberal attitudes toward euthanasia	3.6 ± 0.7	3.5 ± 0.7	0.39
Conservative attitudes toward euthanasia	2.9 ± 0.6	2.9 ± 0.7	0.43
Self-esteem	3.3 ± 1.5	3.4 ± 1.1	0.87
Negative emotions and attitudes	2.1 ± 0.6	2.1 ± 0.7	0.81
Negative thoughts about oneself	1.8 ± 0.6	1.7 ± 0.6	0.68
Unconscious fear of death – associations	1.7 ± 1.5	1.6 ± 1.3	0.76

Data are means ±SD of score points

Multiple regression analysis further indicated that global self-esteem was a positive predictor of conservative attitudes (B = 0.29, t(92) = 2.19; $p < 0.05$), while liberal attitudes were predicted by negative emotions and motivations (B = 0.79, t(92) = 4.38; $p < 0.001$) and by a low intensity of negative thoughts about the self (B = −0.70, t(92) = −3.91; $p < 0.001$).

Correlation analysis was conducted to explore the relationship between unconscious fear of death and attitudes toward euthanasia. The analysis demonstrated a statistically significant, albeit weak, relations between the informational support scale and the association scale (r = 0.25; $p < 0.05$) and the liberal attitude scale and the association scale (r = 0.25; $p < 0.05$).

Cluster analysis using the Ward method was conducted to provide a more complete picture of the relationship between attitudes toward euthanasia and death anxiety dimensions. The resulting dendrogram shows that conservative attitudes form a cluster with unconscious fear of death, while liberal attitudes form clusters with informational support and declared (conscious) fear (Fig. 2).

4   Discussion

In the present study, I investigated several working hypotheses concerning the relationship between attitudes toward euthanasia and the psychological dimensions, such as fear of death and fear of dying, death-thought accessibility, negative mood, and self-esteem. The leading hypothesis presumed that nurses would demonstrate a lower magnitude of fear of death and dying than the lay women in the control group. This hypothesis was confirmed only in relation to the global scores in the Fear of Death and Dying Inventory.

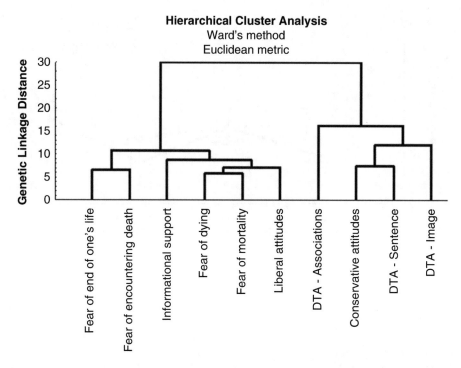

Fig. 2 Genetic linkage distance concerning fear and attitudes toward death and dying; *DTA* death-thought accessibility

Individual scales of the questionnaire failed to show any appreciable differences. The corollary is that nurses working in outpatient clinics or nonsurgical departments, who were not directly exposed to dying patients and death in their professional practice, did not generally differ from the lay women with respect to fear of the end of one's life, mortality, encountering death, or dying. Hence, there is no evidence that practicing a medical profession affects various types of fear of death and dying, which is basically in line with previous findings (Głębocka 2017).

The second hypothesis that nurses would be more liberal and less conservative in their attitudes toward euthanasia and more appreciative of informational support than lay women was not confirmed. In contradistinction, earlier research has led to a conclusion that health-care professionals are less conservative than the lay public concerning the views on euthanasia. The discrepancy may be explicable by the fact that those previous studies involved physicians, nurses, or physiotherapists in contact with the terminally ill (Głębocka 2017), which was not

the case in the current study. Hence, not the workplace itself but rather its specifics seem to determine attitudes toward euthanasia.

The next hypothesis that fear of dying would be predictive of liberal attitudes was not confirmed either. Fear of dying does not increase the acceptance of euthanasia, although it reduces conservative attitudes. People experiencing fear of dying are not so firm and adamant concerning the view that it is unacceptable for patients facing terminal illness to take their own lives. According to the TMT, however, accessibility of death anxiety makes people even more determined to defend commonly accepted views. Perhaps, participants were not fully convinced about the value and validity of their own judgments on non-admissibility of euthanasia. A distinction between conscious (declared) fear, which is the case here, and death-thought accessibility, which is unconscious fear, referred to in TMT, may contribute to the explanation of this result (Rosenblatt et al. 1989; Arndt et al. 1997).

Likewise, the following hypothesis concerning the predictive role of fear of death for conservative attitudes was not confirmed. The findings fail to demonstrate that declared fear of death determined attitudes toward euthanasia in any way. However, this is true for fear of dying, which not only reduces the magnitude of conservative views but also enhances the need for informational and psychological support from physicians and nurses. This result demonstrates that separation of fear of death from fear of dying, as proposed by Wittkowski (2001), is fully justified. Fear of death and fear of dying, though fairly strongly correlated, are distinctively different affective states. Death anxiety is fear of a condition, whereas dying anxiety is fear of a process. Fear of death is abstract in nature and related to nonexistence or potential transition to another unknown reality. Thus, it is accompanied by uncertainty. In contrast, fear of dying involves anxiety about experiencing something known and tangible, namely, pain, suffering, disability, dependence on others, and a loss of control over one's own life.

Subsequent hypotheses directly reflected the TMT proposals. It was expected that death-thought accessibility, measured with projection methods, would enhance a conservative stance. This mechanism involves receiving signals which trigger the subliminal fears. The stronger the anxiety, the easier is to activate it. On the one hand, TMT assumes that death-thought accessibility makes people increasingly supportive for the views that they already consider valid. In this case, liberal persons should become even more liberal, while conservatives should become more conservative. On the other hand, in case of activation of death-related thoughts, people, in an attempt to effectively mitigate their anxiety, turn to simple and unambiguous constructs shared by the general public. Conservative attitudes fulfil these criteria better than liberal ones do. This effect seems obvious, when we consider that the Polish society is deemed conservative, traditional, and highly religious (Czapiński and Panek 2015). The results of the present study contradict the assumptions of the TMT. It turns out that unconscious fear of death was not predictive of conservative attitudes. The study revealed only a weak relationship between indicators of unconscious

fear of death and the need for informational support and for liberal attitude. This effect is not easily explainable. The research procedures or tools might have failed; or it is actually true that unconscious fear of death does not affect one's views on admissibility of euthanasia. Further research of experimental nature, accounting for hidden fear of dying, will be therefore appropriate to resolve these doubts.

Finally, the last hypothesis proposed that negative mood (emotions and motivations) would intensify liberal attitudes. It seems logical and almost certain that negative affect (sadness, dejection) as well as pessimistic vision of the future and a lack of hope for improvement, which are depressive symptoms, may influence a decision to end one's life through euthanasia (Villavicencio-Chávez et al. 2014). Is it equally true that mood determines one's views about euthanasia? The present findings indicate that attitudes accepting euthanasia are predicted by negative emotions and motivations as well as by positive perception of oneself. At this point, it is hard to speculate what psychological mechanism is responsible for the aforementioned relations. Perhaps individuals of high self-esteem, who for some reason experience diminished mood and expect some important matters to go wrong, have lost the sense of causation due to a crisis situation they have been going through. As demonstrated by earlier studies, internal locus of control correlates with a lack of acceptance of euthanasia (Hains and Hulbert-Williams 2013). On the other hand, when internal control is lost, a conviction that euthanasia is admissible might reflect one's attempt to regain the feeling of autonomy and self-determination (Terkamo-Moisio et al. 2017).

When the present findings are confronted with the results of cluster analysis, one may attempt to develop some models of conservative and liberal attitudes toward euthanasia. Conservative attitudes of no euthanasia are partly related to, and partly resulting from, hidden fear of death and high self-esteem. The lack of acceptance of euthanasia, i.e., avoidance of confrontation with death, may be a defense mechanism to manage anxiety. Also, high self-esteem plays a protective role. It is manifest in a sense of one's uniqueness and worth ("I am better than others"). According to Lerner's (1980) just world concept, people believe that everybody gets what he deserves: good people get good things, while bad people get bad outcomes. Consequently, people tend to believe that they will never suffer an incurable disease or slow agony. Liberal attitudes are related to conscious fear and rational vision of oneself, the world, and the future. In particular, they manifest in a conviction that pain and suffering do not necessarily happen only to bad people, but equally often affect good ones, and that it is beyond human being's control when and how they will be dying. Acceptance of euthanasia places the ability to decide in the individual's hands. In such case, conservative attitudes could be considered emotional, whereas liberal ones would be of a rational nature.

Further exploration is required to determine whether such a model is valid. It should be carried out in a wider population, including men, who display higher death anxiety and death-thought accessibility than women do. The present study cannot be considered representative for the whole community of health-care professionals and even less so for the entire population of Poland. This is simply another attempt to describe psychological determinants of attitudes toward euthanasia, which, owing to the ethical aspect of the issue as well as limited availability and peculiar character of study groups (elderly or hospitalized persons), is not an easy task.

Acknowledgments The study was co-financed from the funds allocated to the statutory activity of the Andrzej Frycz Modrzewski Cracow University Faculty of Psychology and Humanities, project ref. WPiNH/DS/1/2017.

Conflicts of Interest The author declares no conflicts of interest in relation to this article.

References

Aghababaei N (2014) Attitudes towards euthanasia in Iran: the role of altruism. J Med Ethics 40(3):173–176

Aghababaei N, Wasserman JA, Hatami J (2014) Personality factors and attitudes toward euthanasia in Iran: implications for end-of-life research and practice. Death Stud 38(1–5):91–99

Arndt J, Greenberg J, Pyszczynski T, Solomon S (1997) Subliminal presentation of death reminders leads to increased defense of the cultural worldview. Psychol Sci 8:379–385

Butcher JN, Hooley JM, Mineka S (2016) Abnormal psychology, 17th edn. Pearson Education Inc. London, UK

Cohen J, Van Landeghem P, Carpentier N, Deliens L (2013) Different trends in euthanasia acceptance across Europe. A study of 13 western and 10 central and eastern European countries, 1981-2008. Eur J Pub Health 23(3):378–380

Collett LJ, Lester D (1969) The fear of death and the fear of dying. J Psychol 72:179–181

Czapiński J, Panek T (2015) Social diagnosis 2015 – objective and subjective quality of life in Poland. http://www.diagnoza.com/pliki/raporty/Diagnoza_raport_2015.pdf. Accessed on 19 Aug 2017

Danyliv A, O'Neill C (2015) Attitudes towards legalising physician provided euthanasia in Britain: the role of religion over time. Soc Sci Med 128(3):52–56

Davidson D, Lymburner JA (2017) Furthering the discussion on a physician-assisted dying right for the mentally ill: commentary on Karesa and McBride (2016). Can Psychol 58(3):292–304

Dymkowski M (1993) The knowledge of oneself. Publishing House of the Institute of Psychology of the Polish Academy of Sciences (Article in Polish), Warsaw

Gesang B (2008) Passive and active euthanasia: what is the difference? Med Health Care Philos 11(2):175–180

Gielen J, van den Branden S, Broeckaert B (2009) Religion and nurses' attitudes to euthanasia and physician assisted suicide. Nurs Ethics 16(3):303–318

Głębocka A (2017) Attitudes towards euthanasia in the context of fear of death among physiotherapists and caregivers of patients with paresis. Med Sci Pulse 3 (11):15–20 (Article in Polish)

Głębocka A, Gawor A, Ostrowski F (2013) Attitudes toward euthanasia among Polish physicians, nurses and people who have no professional experience with the terminally ill. Adv Exp Med Biol 788:407–412

Goryńska E (2005) The mood adjective checklist (UMACL) by Matthews, Chamberlain and Jones. PTP, Warsaw

Greenberg J, Solomon S, Pyszczynski T, Rosenblatt A, Burling J, Lyon D, Pinel E, Simon L (1992) Assessing the terror management analysis of self-esteem: converging evidence of an anxiety-buffering function. J Pers Soc Psychol 63:913–922

Hains CM, Hulbert-Williams NJ (2013) Attitudes toward euthanasia and physician-assisted suicide: a study of the multivariate effects of healthcare training, patient characteristics, religion and locus of control. J Med Ethics 39(11):713–716

Haque OS, Waytz A (2012) Dehumanization in medicine: causes, solutions, and functions. Perspect Psychol Sci 7(2):176–186

Ho R, Penney RK (1992) Euthanasia and abortion: personality correlates for the decision to terminate life. J Soc Psychol 132(1):77–86

Kirmes T, Wilk M, Chowaniec C (2017) Euthanasia – an attempt to organize issue. Wiadomosci Lekarskie 70 (1):118–127

Kneer J, Rieger D (2016) The memory remains: how heavy metal fans buffer against the fear of death. Psychol Pop Media Cult 5(3):258–272

Lerner MJ (1980) The belief in a just world. In: The belief in a just world. Perspectives in social psychology. Springer, Boston

Matthews G, Jones DM, Chamberlain AG (1990) Refining the measurement of mood: the UWIST mood adjective checklist. Br J Psychol 81:17–42

Ochsmann R (1984) Belief in afterlife as a moderator of fear of death? Eur J Soc Psychol 14(1):53–67

Pyszczynski T, Greenberg J, Solomon S (1999) A dual-process model of defense against conscious and unconscious death-related thoughts: an extension of terror management theory. Psych Rev 106(4):835–845

Roelands M, Vanden Block L, Geurts S, Deliens L, Cohen J (2015) Attitudes of Belgian students of medicine, philosophy and law toward euthanasia and the conditions for its acceptance. Death Stud 39 (3):139–150

Rosenblatt A, Grennberg J, Solomon S, Pyszczynski T, Lyon D (1989) Evidence for terror management theory: the effects of mortality salience on reactions to those who violate or uphold cultural values. J Pers Soc Psychol 57(4):681–690

Śmieja M, Kałaska M, Adamczyk M (2006) Scared to death or scared to love? Terror management theory and close relationships seeking. Eur J Soc Psychol 36:279–296

Terkamo-Moisio A, Kvist T, Laitila T, Kangasniemi M, Ryynanen O, Pietilä A (2017) The traditional model does not explain attitudes toward euthanasia: a web-based survey of the general public in Finland. Omega (Westport) 75(3):266–283

Vaes J, Muratore M (2013) Defensive dehumanization in the medical practice: a cross-sectional study from a health care worker's perspective. Brit J Soc Psychol 52:180–190

Villavicencio-Chávez C, Monforte-Royo C, Tomás-Sábad J, Maier MA, Porta-Sales J, Balaguer A (2014) Physical and psychological factors and the wish to hasten death in advanced cancer patients. Psycho-Oncology 23(10):1125–1132

Wittkowski J (2001) The construction of the multidimensional orientation toward dying and death inventory. Death Stud 11:479–495

Adv Exp Med Biol - Clinical and Experimental Biomedicine (2018) 1: 105–109
https://doi.org/10.1007/5584_2018_196
© Springer International Publishing AG, part of Springer Nature 2018
Published online: 7 April 2018

Relation Between Attention-Deficit Hyperactivity Disorder and IgE-Dependent Allergy in Pediatric Patients

Mateusz Miłosz, Urszula Demkow, and Tomasz Wolańczyk

Abstract

Food allergy is a common condition in children and adolescent, remitting with time. Few clinical studies have emphasized the link between food allergies and psychosocial conditions, suggesting a profound impact of atopic diseases on the development of attention-deficit hyperactivity disorder (ADHD) in children. The objective of this study was to compile and assess available studies on the comorbidity or causality between ADHD and atopic food allergy in children. We discuss epidemiology, interrelated mechanisms, and potential dietary interventions in the management of children with ADHD.

Keywords

ADHD · Allergy · Atopy · Children · IgE · Food · Psychosocial status

1 Attention-Deficit Hyperactivity Disorder (ADHD)

Attention-deficit hyperactivity disorder (ADHD) is a neurodevelopmental disorder in which first manifestations occur in childhood (Bierderman and Faraone 2005). The main symptoms are hyperactivity, impulsivity, and attention deficit, which are inappropriate for the child's age and are present across a range of settings causing impairment in functioning. The prevalence of ADHD is about 5% in children and adolescents and 3% in adults (Polanczyk et al. 2007; Swanson et al. 1998). ADHD is believed to equally affect children of all social classes. However, strong evidence supports the notion that ADHD is more common among the poor (Russell 2016).

There are three main types of the disease according to the Diagnostic and Statistical Manual of Mental Disorders (American Psychiatric Association 2013): having the predominance of hyperactivity and impulsivity, or attention deficit, or a mixture of both. ADHD is highly heritable. A study performed in twins reports the mean heritability of ADHD to be around 76%. On the other hand, biopsychosocial models of ADHD include both genetic and environmental, including epigenetic, inflammatory, toxic, social, and other interactions leading to increased risk of ADHD. Clearly, there is no simple causal explanation (Kollins et al. 2008; Vaidya and Stollstorff 2008; Kim et al. 2006). Whether ADHD and

M. Miłosz (✉) and U. Demkow
Department of Laboratory Diagnostics and Clinical Immunology of Developmental Age, Warsaw Medical University, Warsaw, Poland
e-mail: m.milosz@euroimmun.pl

T. Wolańczyk
Deaprtment of Children and Adolescent Psychiatry, Warsaw Medical University, Warsaw, Poland

allergy are interrelated is still an open and highly disputable question. Reports concerning allergies in children with ADHD, suggesting a possible causal background, have created a growing concerning among patients and families (Lin et al. 2016).

2   Allergy

Food allergy is a process that occurs reproducibly after intake of certain foods. The process includes an immunologic response with IgE production or it may be non-IgE-mediated. It can also be of non-immunologic background, assuming the form of intolerance, the exemplary of which may be lactose intolerance. The IgE-mediated food allergy can be confirmed by allergen-specific IgE testing and by specific clinical symptoms occurring after exposure to allergens (Rona et al. 2007).

Clinical symptoms of IgE-dependent allergy include urticarial problems, angioedema, cough, runny nose, vomiting, headache, severe cardiac complications, and anaphylaxis (Schnyder and Pichler 2009). This type of reactions, characterized by the production of specific IgE antibodies, is the most common. Symptoms occur shortly after contact with allergen. Allergens enter the body from the air or ingested food (Sampson 1999). In industrialized countries, more than 15% of people suffer for direct allergy, resulting in rhinitis, asthma, conjunctivitis, or atopic dermatitis. Food allergies, which involve the formation of IgE antibodies, give rise to symptoms hours after food digestion. The symptoms include burning, itching, nausea and vomiting, abdominal pain in extreme cases, asthma, confusion, and even anaphylaxis. These symptoms often appear after ingestion of peanuts and fish (Ghunaim et al. 2005).

Pediatrics patients with food allergy are 2–4 times more likely to have related conditions, such as atopic dermatitis, asthma, or allergic rhinitis, compared to children without food allergy (Branum and Lukacs 2009). A risk for developing food allergy associates with preexisting allergic disease or family history of food allergy (Rona

et al. 2007). There are currently no recommendations on the medications for preventing IgE- or non-IgE-mediated food-induced allergy reactions (Boyce et al. 2010). The most certain method to avoid allergic reactions is having a diet free of specific allergens. Children with food allergy and their caregivers should be instructed in the interpretation of food labels. Another possibility is to retest the food suspected of causing an allergic response, at a time interval that depends on the child's age and medical history (Sampson 1999).

The presence of IgE is usually confirmed using an in vivo prick skin test or in vitro tests assessing the level of IgE in the serum (Burks et al. 2011). Skin testing have some limitations, e.g., depending on the skin condition, which underscores the role IgE testing in the serum (Sicherer et al. 2012). Currently, single allergen testing, using a fluoroluminescence method, has been accepted as a reference method for specific IgE diagnostics. A disadvantage of the method is the possibility of only a single allergen detection, which may impede a complete assessment of a disease and increase the cost of further diagnostic efforts. Multiple allergen simultaneous testing (MAST) is more profitable to this end (Shin et al. 2010).

3   Allergy in Patients with Attention-Deficit Hyperactivity Disorder

Feingold (1975) has proposed a hypothesis that different types of food that could act as allergens in atopic diseases may also lead to the development of hyperactivity. Coexistence of allergy in patients with ADHD has been studied since the early 1980s. The studies have clearly noted an increase in the prevalence of allergic disease accompanying ADHD. However, majority of data are anecdotal and of low quality (Miyazaki et al. 2017; Belfer 2008). The relationship between allergy and ADHD may be divided into two subgroups: one focusing on behavioral symptoms, including symptoms of ADHD in children with allergic diseases, and the other that is more epidemiologically oriented and focuses on comorbidities. Patients with allergic disease such as eczema, asthma, or allergic rhinitis may also exhibit hyperactive and impulsive behavior. These patients may need an integrated combination of diagnostics, prevention, and treatment strategies (Boris and Mandel 1994). For instance, scores on the hyperactivity/impulsivity and inattention subscales are significantly higher in children with allergic rhinitis than in control subjects, although such children may have not necessarily been diagnosed with ADHD (Feng et al. 2017).

Schmitt et al. (2009) have studied a group of 1436 patients with atopic dermatitis and found direct relationship to the co-occurrence of ADHD. Fasmer et al. (2011) have reported an increased prevalence of asthma in ADHD patients, and those who have more symptoms of ADHD have a still higher rate of asthma. Similar conclusions have been drawn by Mogensen et al. (2011) in a study involving 1480 pairs of twins. Chen et al. (2013) have gathered information the allergic diseases coexisting with ADHD and found that 25.2% of ADHD patients have asthma, 40.6% have allergic rhinitis, and 17.9% have atopic dermatitis. Chou et al. (2013) have found that ADHD patients have a significantly higher prevalence of allergic rhinitis (28.4%) compared with the general population (15.2%). The prevalence of allergic disease in ADHD patients is independent of environmental and lifestyle factors such as parental smoking, breast feeding, or the beginning of day care at an early age (Schmitt et al. 2010).

Food hypersensitivity, including true allergy, and ADHD may share etiologic pathways, as a behavioral response to food often occurs in ADHD. There are studies that evaluate the effect of dietary restrictions and point to significant benefits gained from the elimination diet in some patients suffering from attention hyperactivity disorders (Verlaet et al. 2014). Chou et al. (2013) have reported that ADHD patients have a higher prevalence of allergy and asthma, but not atopic dermatitis, than the general population has. In addition, ADHD is more prevalent in boys than girls and often is diagnosed when children are at school. The prevalence of ADHD also is greater

in the urban areas. In the US study conducted in children and adolescents, which encompassed 354,416 subjects, Strom et al. (2016) have reported that the severity of atopic dermatitis is associated with that of ADHD. The prevalence of ADHD also was overall greater in children with the accompanying allergic diseases, and it was estimated at 7.3% in eczema, 10.9% in asthma, 8.9% in hay fever, 11.9% in eczema and asthma, 9.8% in eczema and hay fever, 12.9% in asthma and hay fever, and 14.5% in eczema and asthma and hay fever. ADHD is definitely a frequent comorbidity of atopic dermatitis, particularly the genetically underlain form of atopic dermatitis linked to the filaggrin gene mutations (Chiesa Fuxench 2017). Likewise, a systematic review and meta-analysis published in 2017 has revealed that children with ADHD are more likely to have asthma, allergic rhinitis, atopic dermatitis, and allergic conjunctivitis than the healthy peers (Miyazaki et al. 2017), but no such relationship was substantiated in case of food allergy. Similar results have been published by Chen et al. (2017) concerning asthma, atopic eczema, and allergic rhinitis. Moreover, the authors demonstrate that individuals with allergic diseases have about 50% greater chance of developing ADHD compared to healthy subjects. A biological plausibility exists that ADHD and allergic disorders share a common background at the translational level involving gene interactions, although the exact mechanisms of this association are unclear as it lacks causality. Nonetheless, molecular biology has provided some possible cues on the association between allergic disorders and ADHD. One is that patients with atopic diseases are exposed to higher levels of pro-inflammatory cytokines which permeate through the blood-brain barrier and activate neuro-immune mechanisms involving emotional and behavioral symptoms related to ADHD (Schmitt et al. 2010). Another possibility is that allergy could impair the mechanisms of synaptic plasticity in the prefrontal cortex, causing cognitive brain dysfunction characteristic of ADHD (Goto et al. 2010). In support of that notion, Trikojat et al. (2015) have found that patients with seasonal allergic rhinitis have a slowdown in information processing speed, which causes changes in attentional control, both in and off symptomatic season, and which associates with IgE levels. Ishiuji et al. (2009) and Sun et al. (2012) have investigated cortical cingulate activity involved with affect and emotion, in atopic dermatitis patients and healthy controls during histamine-induced itch, using functional magnetic resonance imaging. The authors demonstrate changes in cortical activity in atopic dermatitis, compared with the healthy condition, which associate with the measures of disease intensity, including impulsivity and inattention.

4   Summary and Conclusions

Epidemiological population studies show an association between allergic disorders and attention-deficit hyperactivity disorder. A variety of allergies associates with different subtypes of ADHD. It has been noticed that the highest incidence of atopic dermatitis occurs in patients with allergic rhinitis and in ADHD patients in whom attention deficit is a predominant symptom. Genetic problems linked to immunity, dysfunction of cortical plasticity, and the influence of environmental factors are the presumed pathological underpinnings of allergies and ADHD, but the exact mechanisms underlying the co-occurrence of both pathologies remain unclear.

The association of allergy and ADHD is of raising clinical interest as both pathologies are sharply on the rise globally and notably involve childhood and adolescent periods of life. The most common diseases that associate with ADHD are atopic dermatitis, allergic rhinitis, and asthma. We believe this brief review will serve to raise awareness of the issue and to indicate the need for meticulous workup to identify the agents one is allergic and to assess cognitive functions.

Conflicts of Interest The authors declare no conflicts of interest in relation to this paper.

References

American Psychiatric Association (2013) Diagnostic and statistical manual of mental disorders, 5th edn. American Psychiatric Publishing, Arlington. ISBN 978-0-89042-555-8

Belfer ML (2008) Child and adolescent mental disorders: the magnitude of the problem across the globe. J Child Psychol Psychiatry Allied Disc 49(3):226–236

Bierderman J, Faraone SV (2005) Attention-deficit hyperactivity disorder. Lancet 366:237–248

Boris M, Mandel FS (1994) Food and additives are common causes of the attention deficit hyperactive disorder in children. Ann Allergy 72(5):462–468

Boyce JA, Assa'ad A, Burks AW et al (2010) Guidelines for the diagnosis and management of food allergy in the United States: report of the NIAID in the United States: report of the NIAID-sponsored expert panel. J Allergy Clin Immunol 126(6):1105–1118

Branum AM, Lukacs SL (2009) Food allergy among children in the United States. Pediatrics 124(6):1549–1555

Burks AW, Jones SM, Boyce JA et al (2011) NIAID-sponsored 2010 guidelines for management of food allergy: applications in the pediatric population. Pediatrics 128:955–965

Chen MH, Su TP, Chen YS, Hsu JW, Huang KL, Chang WH, Bai YM (2013) Allergic rhinitis in adolescence increases the risk of depression in later life: a nationwide population-based prospective cohort study. J Affect Disord 145(1):49–53

Chen MH, Su TP, Chen YS, Hsu JW, Huang KL, Chang WH, Chen TJ, Bai YM (2017) Comorbidity of allergic and autoimmune diseases among patients with ADHD. J Atten Disord 21(3):219–227

Chiesa Fuxench ZC (2017) Atopic dermatitis: disease background and risk factors. Adv Exp Med Biol 1 (027):11–19

Chou PH, Lin CC, Lin CH, Loh e-W, Chan CH, Lan TH (2013) Prevalence of allergic rhinitis in patients with attention-deficit/hyperactivity disorder: a population-based study. Eur Child Adolesc Psychiatry 22:301–307

Fasmer OB, Halmøy A, Eagan TM, Oedegaard KJ, Haavik J (2011) Adult attention deficit hyperactivity disorder is associated with asthma. BMC Psychiatry 11:128

Feingold BF (1975) Hyperkinesis and learning disabilities linked to artificial food flavors and colors. Am J Nurs 75:797–803

Feng B, Jin H, Xiang H, Li B, Zheng X, Chen R, Shi Y, Chen S, Chen B (2017) Association of pediatric allergic rhinitis with the ratings of attention-deficit/hyperactivity disorder. Am J Rhinol Allergy 31 (3):161–167

Ghunaim N, Grönlund H, Kronqvist M, Grönneberg R, Söderström L, Ahlstedt S, van Hage-Hamsten M (2005) Antibody profiles and self-reported symptoms to pollen-related food allergens in grass pollen-allergic patients from northern Europe. Allergy 60(2):185–191

Goto Y, Yang CR, Otani S (2010) Functional and dysfunctional synaptic plasticity in prefrontal cortex: roles in psychiatric disorders. Biol Psychiatry 67 (3):199–207

Ishiuji Y, Coghill RC, Patel TS, Oshiro Y, Kraft RA, Yosipovitch G (2009) Distinct patterns of brain activity evoked by histamine-induced itch reveal an association with itch intensity and disease severity in atopic dermatitis. Br J Dermatol 161(5):1072–1080

Kim JW, Kim BN, Cho SC (2006) The dopamine transporter gene and the impulsivity phenotype in attention deficit hyperactivity disorder: a case-control association study in a Korean sample. J Psychiatr Res 40 (8):730–737

Kollins SH, Anastopoulos AD, Lachiewicz AM, FitzGerald D, Morrissey-Kane E, Garrett ME, Keatts SL, Ashley-Koch AE (2008) SNPs in dopamine D2 receptor gene (DRD2) and norepinephrine transporter gene (NET) are associated with continuous performance task (CPT) phenotypes in ADHD children and their families. Am J Med Genet B Neuropsychiatr Genet 147B(8):1580–1588

Lin YT, Chen YC, Gau SS, Yeh TH, Fan HY, Hwang YY, Lee YL (2016) Associations between allergic diseases and attention deficit hyperactivity/oppositional defiant disorders in children. Pediatr Res 80(4):480–485

Miyazaki C, Koyama M, Ota E, Swa T, Mlunde LB, Amiya RM, Tachibana Y, Yamamoto-Hanada K, Mori R (2017) Allergic diseases in children with attention deficit hyperactivity disorder: a systematic review and meta-analysis. BMC Psychiatry 17:120

Mogensen N, Larsson H, Lundholm C, Almqvist C (2011) Association between childhood asthma and ADHD symptoms in adolescence: a prospective population-based twin study. Allergy 1:224–1230

Polanczyk G, de Lima MS, Horta BL, Biederman J, Rohde LA (2007) The worldwide prevalence of ADHD: a systematic review and meta-regression analysis. Am J Psychiatry 164(6):942–948

Rona RJ, Keil T, Summers C, Gislason D, Zuidmeer L, Sodergren E, Sigurdardottir ST, Lindner T, Goldhahn K, Dahlstrom J, McBride D, Madsen C (2007) The prevalence of food allergy: a meta-analysis. J Allergy Clin Immunol 120(3):638–646

Russell AE (2016) The association between socioeconomic disadvantage and attention deficit/hyperactivity disorder (ADHD): a systematic review. Child Psychiatry Hum Dev 47:440–458

Sampson HA (1999) Food allergy. Part 1: immunopathogenesis and clinical disorders. J Allergy Clin Immunol 103(5 pt 1):717–728

Schmitt J, Romanos M, Schmitt NM, Meurer M, Kirch W (2009) Atopic eczema and attention-deficit/hyperactivity disorder in a population-based sample of children and adolescents. JAMA 7:724–726

Schmitt J, Buske-Kirschbaum A, Roessner V (2010) Is atopic disease a risk factor for attention-deficit/hyperactivity disorder? A systematic review. Allergy 65(12):1506–1524

Schnyder B, Pichler WJ (2009) Mechanisms of drug-induced allergy. Mayo Clin Proc 84(3):268–272

Shin JW, Jin S, Lee JH, Cho S (2010) Analysis of MAST-CLA results as a diagnostic tool in allergic skin diseases. Ann Dermatol 22:35–40

Sicherer SH, Wood RA, American Academy of Pediatrics Section on Allergy and Immunology (2012) Allergy testing in childhood: using allergen-specific IgE tests. Pediatrics 129(1):193–197

Strom MA, Fishbein AB, Paller AS, Silverberg JI (2016) Association between atopic dermatitis and attention deficit hyperactivity disorder in U.S. children and adults. Br J Dermatol 175(5):920–929

Sun L, Cao Q, Long X, Sui M, Cao X, Zhu C, Zuo X, An L, Song Y, Zang Y, Wang Y (2012) Abnormal functional connectivity between the anterior cingulate and the default mode network in drug-naive boys with attention deficit hyperactivity disorder. Psychiatry Res 201(2):120–127

Swanson J, Castellanos FX, Murias M, LaHoste G, Kennedy J (1998) Cognitive neuroscience of attention deficit hyperactivity. Curr Opin Neurobiol 8 (2):263–271

Trikojat K, Buske-Kirschbaum A, Schmitt J, Plessow F (2015) Altered performance in attention tasks in patients with seasonal allergic rhinitis: seasonal dependency and association with disease characteristics. Psychol Med 45(6):1289–1299

Vaidya CJ, Stollstorff M (2008) Cognitive neuroscience of attention deficit hyperactivity disorder: current status and working hypotheses. Dev Disabil Res Rev 117 (18):2407–2423

Verlaet AA, Noriega DB, Hermans N, Savelkoul HF (2014) Nutrition, immunological mechanisms and dietary. Eur Child Adolesc Psychiatry 23(7):519–529